Springer Theses

Recognizing Outstanding Ph.D. Research

Aims and Scope

The series "Springer Theses" brings together a selection of the very best Ph.D. theses from around the world and across the physical sciences. Nominated and endorsed by two recognized specialists, each published volume has been selected for its scientific excellence and the high impact of its contents for the pertinent field of research. For greater accessibility to non-specialists, the published versions include an extended introduction, as well as a foreword by the student's supervisor explaining the special relevance of the work for the field. As a whole, the series will provide a valuable resource both for newcomers to the research fields described, and for other scientists seeking detailed background information on special questions. Finally, it provides an accredited documentation of the valuable contributions made by today's younger generation of scientists.

Theses are accepted into the series by invited nomination only and must fulfill all of the following criteria

- They must be written in good English.
- The topic should fall within the confines of Chemistry, Physics, Earth Sciences, Engineering and related interdisciplinary fields such as Materials, Nanoscience, Chemical Engineering, Complex Systems and Biophysics.
- The work reported in the thesis must represent a significant scientific advance.
- If the thesis includes previously published material, permission to reproduce this must be gained from the respective copyright holder.
- They must have been examined and passed during the 12 months prior to nomination.
- Each thesis should include a foreword by the supervisor outlining the significance of its content.
- The theses should have a clearly defined structure including an introduction accessible to scientists not expert in that particular field.

More information about this series at http://www.springer.com/series/8790

Tjonnie G.F. Li

Extracting Physics from Gravitational Waves

Testing the Strong-field Dynamics of General Relativity and Inferring the Large-scale Structure of the Universe

Doctoral Theses accepted by
VU University Amsterdam, The Netherlands

Springer

Author
Dr. Tjonnie G.F. Li
LIGO Laboratory
California Institute of Technology
Pasadena, CA
USA

Supervisor
Dr. Chris F.F. Van Den Broeck
National Institute for Subatomic Physics—Nikhef
Amsterdam
The Netherlands

ISSN 2190-5053 ISSN 2190-5061 (electronic)
Springer Theses
ISBN 978-3-319-36664-7 ISBN 978-3-319-19273-4 (eBook)
DOI 10.1007/978-3-319-19273-4

Springer Cham Heidelberg New York Dordrecht London

Softcover reprint of the hardcover 1st edition 2015

Printed on acid-free paper

Springer International Publishing AG Switzerland is part of Springer Science+Business Media (www.springer.com)

「學而時習之，不亦説乎？」
"Isn't it a pleasure to study and practice what you have learned?"
孔子/ Confucius (551–479 BC)

Supervisor's Foreword

At its 100th anniversary, Einstein's theory of general relativity (GR) still encapsulates our best understanding of space, time, and gravitation. Early tests of the framework included the perihelium precession of Mercury, the deflection of starlight by the Sun, and Shapiro time delay. Although excellent agreement with theory was obtained, these checks only probed the effect on the motion of test masses of low-order general-relativistic corrections to the gravitational field. The situation improved dramatically with the discovery of the Hulse-Taylor binary neutron star in 1974, which confirmed the existence of gravitational radiation at leading-order (quadrupole) level. Subsequently, more relativistic binary systems were discovered, allowing for impressive new tests of GR. Nevertheless, if one is interested in further probing the dissipative dynamics of binaries, and especially the dynamics of spacetime itself, what matters is the orbital compactness $GM/(c^2R)$ (with M the total mass and R the separation), as well as the orbital velocity v/c. Even the recently discovered neutron star-white dwarf system only has $GM/(c^2R) \sim 2 \times 10^{-6}$, and $v/c) \sim 4 \times 10^{-3}$.

By contrast, binaries consisting of neutron stars and/or black holes on the verge of merger will have $GM/(c^2R) > 0.2$ and $v/c > 0.4$, with copious gravitational wave (GW) emission. The ability to observe the orbital motion of such systems would give us access to the genuinely strong-field, relativistic regime of gravity. Most importantly, we would like to probe the dynamical self-interaction of spacetime itself, such as the scattering of gravitational waves off the spacetime curvature generated by the binary as a whole. The only way to gain empirical access to such phenomena is through the direct detection of gravitational waves. A network of advanced interferometric gravitational wave detectors is currently under construction, including the Advanced LIGO in the US, Advanced Virgo in Italy, and KAGRA in Japan; GEO-HF in Germany is already active, and an Advanced LIGO interferometer may be placed in India. Towards the end of the decade, these observatories may find tens of GW signals per year from coalescing compact binaries.

There is a considerable body of literature on (families of) alternative theories of gravity: models with an anomalous dispersion relation for the "graviton", scalar-tensor theories, parity-violating theories, models with higher curvature corrections, and more. Using GW signals from coalescing binaries to look for evidence for each of these separately might be impractical; moreover, if a GR violation is present then it may well be of a kind that has not yet been envisaged. For that reason, it is important to test GR in a *model-independent* way, by developing a method that is capable of uncovering the effects of a very wide range of ways in which GR might break down. Moreover, in order to be useful in the immediate future, the method should be able to deal with the relative quiet coalescence signals that we expect to see with the GW observatories currently under construction. Ideally, it would also allow for combining information from multiple detections in order to arrive at a statement about the validity of GR (or lack thereof) that is as strong as possible.

Such a framework is introduced in this book. This Bayesian model selection scheme satisfies the desiderata listed above, and has since been shown to be robust against a number of nuisance effects of a fundamental, astrophysical, and instrumental nature, at least in the case of coalescing binary neutron stars; extending it to GW events involving black holes is currently in progress. Thus, the method presented here is poised to become a standard tool for empirical study of the strong-field dynamics of spacetime.

Another field that will greatly benefit from the direct detection of GW signals from coalescing binaries is cosmology. This is because coalescence events are "standard sirens", similar to the electromagnetic (EM) "standard candles" such as Type Ia supernovae. However, an important difference between EM and GW cosmic distance markers is that the latter are *self-calibrating*: the luminosity distance to binary coalescences can be inferred from the GW signals alone, without having to compare with any other kind of source, thus obviating the need for a cosmic distance ladder and avoiding the systematic errors that may be associated with it. Precision measurement of the Hubble constant will already be possible with the upcoming advanced gravitational wave detectors. Meanwhile, a conceptual design study has been performed for a third-generation observatory called Einstein Telescope, which will be capable of seeing about three orders of magnitude more coalescence events per year, out to high redshifts. As shown in this book, it will be possible to use the Einstein Telescope to measure the densities of matter and dark energy, as well as the dark energy equation of state and its possible time dependence, with accuracies comparable to those of future EM surveys, but in a completely independent way.

The subject matter of this book is at the forefront of the nascent fields of observational gravitational wave physics and cosmology. At the same time the discussion is to a large degree self-contained, including an accessible introduction to the theory behind gravitational wave emission and waveform modelling, as well as state-of-the-art data analysis methods. As such it will be an invaluable resource for young researchers as we enter an exciting new era in the study of gravitation.

Amsterdam, The Netherlands
April 2015

Dr. Chris F.F. Van Den Broeck

Preface

Humans have been watching the sky for thousands of years. In early times, humans tracked the motion of the Sun and the Moon to make calendars and to associate it with earthly events such as tides and seasons. By tracking the motion of celestial objects, the early notion of the orbit of the Sun, the Moon, and the planets started to form. Isaac Newton (1642–1727) showed that all objects "pull" on each other through gravitational force, and the strength of this force is determined by the masses of the two objects.

In 1915, Albert Einstein (1879–1955) refined Newton's law of universal gravitation by introducing the general theory of relativity or general relativity for short. General relativity describes how mass distorts spacetime and, in turn, how spacetime dictates how masses flow through it. Moreover, general relativity predicts that gravity is mediated by a new type of radiation: *gravitational radiation.* Gravitational waves that compose gravitational radiation are ripples in the fabric of spacetime, which periodically lengthen and shorten space, and speed up and slow down time.

However, it was not until 1974 that the effects of gravitational radiation were first measured, albeit indirectly, by Hulse and Taylor. The changes in the orbital motion of two pulsars were remarkably consistent with the emission of gravitational radiation. Despite this tremendous discovery, which was awarded the 1993 Nobel Prize in physics, we have yet to *directly* detect minor distortions of spacetime caused by passing gravitational waves.

Large-scale physics experiments such as the United States-based Laser Interferometer Gravitational Wave Observatory (LIGO) and the Italy-based Virgo aim to, for the first time in the history of mankind, detect the influences of gravitational waves directly. These detectors are set up to measure tiny changes in distances of about one thousandth the diameter of a proton. In 2010, both LIGO and Virgo were decommissioned to make way for significantly improved versions of these detectors, the so-called Advanced LIGO and Advanced Virgo. The Advanced detectors are scheduled to be completed in 2015, and scientists are optimistic that the first direct detection of gravitational waves is imminent.

But the prospects of the first direct detection of gravitational waves extend much beyond the confirmation of Einstein's predictions. In fact, the first direct detection of gravitational waves promises to usher in a new era of astrophysics and astronomy. In particular, measurements of gravitational waves directly probe gravity, the interaction experienced by all known forms of matter, even those that do not emit electromagnetic radiation. Therefore, measurements of gravitational waves could not only supplement our existing knowledge, they could also reveal yet unknown astrophysical phenomena that could radically change our view of the Universe. Indeed, direct measurement of gravitational waves promises to open a new window into the Universe.

For example, one can use gravitational-wave measurements to test general relativity. History has taught us that our laws of physics need continual refinement to describe Nature to ever-increasing accuracy, and no exception should be made for Einstein's general theory of relativity. In fact, Einstein's theory has never been tested in the vicinity of extremely dense objects such as neutron stars and black holes. The gravitational-wave sources that Advanced LIGO and Virgo are expected to see will give us access to the behaviour of gravity in these extreme environments, and allow us to test general relativity with unprecedented accuracy.

Moreover, gravitational-wave measurements provides a way to map distance and motion of celestial objects and study their large-scale behaviour. Existing measurements suggest that the Universe is undergoing an accelerated expansion. This can be described by general relativity, but at the expense of introducing dark matter and dark energy, which combined makes up more than 90 % of the content of the Universe. Gravitational-wave measurements allow us to test this paradigm independently and possibly probe these unknown entities that appear to dominate our Universe.

Indeed, it is quintessential to look beyond the detection of gravitational waves, and to think about how one can utilise gravitational-wave measurements to study the nature of gravity and the astrophysical processes it governs. To achieve this, the gravitational-wave community needs to develop increasingly accurate models of the sources of gravitational waves, to prepare the data-analysis tools to extract relevant information, and finally to use the newly acquired insight to complement or to alter our current state of knowledge. The author hopes that the work described in this book will be a step in the right direction.

This book is divided into three parts. Part I serves as a general introduction to the field of gravitational-wave data analysis. Part II introduces the Test Infrastructure for GEneral Relativity (TIGER), a data-analysis pipeline designed to test general relativity without the need for knowledge of a specific alternative. Finally, Part III focusses on inferring the large-scale structure of the Universe from gravitational-wave signals emitted by merging neutron stars.

Lastly, the author wishes to thank his colleagues at Nikhef, the California Institute of Technology, the LIGO Scientific Collaboration, and the Virgo Collaboration for embarking on the mission of measuring gravitational waves with him. Moreover, the author wants to express gratitude to his parents, brother, and Ivy for their continual support.

Pasadena, USA
April 2015

Dr. Tjonnie G.F. Li

Acknowledgments

This work was carried out at Nikhef as part of the research programme of the "Stichting voor Fundamenteel Onderzoek der Materie (FOM)", which is financially supported by the "Nederlandse Organisatie voor Wetenschappelijk Onderzoek (NWO)".

Contents

Notations and Conventions

Units

Unless explicitly stated, we will use the geometrised units for which the velocity of light c and Newton's constant G are both set to unity, *i.e.*

$$c = G = 1. \tag{1}$$

Signature Conventions and Notations

Lower-case Greek letters denote spacetime indices and take the values of $0, \ldots, 3$. On the other hand, lower-case Roman letters denote space indices and take the values of $1, \ldots, 3$. A spacetime event x^μ is given by

$$x^\mu = (t, x^i) \quad \text{for } \mu = 0, 1, 2, 3 \text{ and } i = 1, 2, 3, \tag{2}$$

where t represents the coordinate time and x^i denote the space coordinates. The signature convention for the metric $g_{\mu\nu}$ is (−,+,+,+). For example, the metric tensor describing flat spacetime in the usual (t, x, y, z) coordinates, is given by

$$\eta_{\mu\nu} = \text{diag}(-1, 1, 1, 1). \tag{3}$$

The invariant line element, which can be constructed from the metric tensor, is given by

$$ds^2 = g_{\mu\nu} dx^\mu dx^\nu. \tag{4}$$

A comma followed by an index denotes a partial derivative, *i.e.*

$$V^{\nu}{}_{,\mu} \equiv \frac{\partial V^{\nu}}{\partial x^{\mu}}. \tag{5}$$

A semi-colon followed by an index denotes the covariant derivative, *i.e.*

$$V^{\nu}{}_{;\mu} \equiv V^{\nu}{}_{,\mu} + V^{\alpha}\Gamma^{\nu}{}_{\alpha\mu}, \tag{6}$$

where $\Gamma^{\rho}{}_{\mu\nu}$ is the Christoffel symbol given by

$$\Gamma^{\rho}{}_{\mu\nu} = \frac{1}{2} g^{\sigma\rho}\left(g_{\sigma\mu,\nu} + g_{\sigma\nu,\mu} - g_{\mu\nu,\sigma}\right). \tag{7}$$

The Riemann tensor is defined as

$$R^{\mu}{}_{\nu\rho\sigma} \equiv \Gamma^{\mu}{}_{\nu\sigma,\rho} - \Gamma^{\mu}{}_{\nu\rho,\sigma} + \Gamma^{\mu}{}_{\alpha\rho}\Gamma^{\alpha}{}_{\nu\sigma} - \Gamma^{\mu}{}_{\alpha\sigma}\Gamma^{\alpha}{}_{\nu\rho}. \tag{8}$$

The Ricci tensor and Ricci scalar are, respectively, given by

$$R_{\mu\nu} \equiv R^{\alpha}{}_{\mu\alpha\nu}, \quad \text{and } R \equiv g^{\mu\nu} R_{\mu\nu}. \tag{9}$$

Acronyms

BAO	Baryon acoustic oscillations
BBH	Binary black hole
BH	Black hole
BHNS	Black hole neutron star binary
BNS	Binary neutron star
CBC	Compact binary coalescence
CMB	Cosmic microwave background
CW	Continuous waves
EFE	Einstein field equations
EM	Electromagnetic
ET	Einstein Telescope
FAP	False alarm probability
FLRW	Friedmann-Lemaître-Robertson-Walker
GR	General relativity
GW	Gravitational wave
LAL	LIGO Algorithms Library
LHS	Left hand side
LIGO	Laser Interferometer Gravitational Wave Observatory
LLF	Local Lorentz frame
LSO	Last stable orbit
MCMC	Markov chain Monte Carlo
NS	Neutron star
PM	Post-Minkowskian
PN	Post-Newtonian
PPE	Parameterised post-Einsteinian
RHS	Right hand side
SDSS	Sloan Digital Sky Survey

SGRB	Short γ-ray burst
SNR	Signal-to-noise ratio
SPA	Stationary phase approximation
STF	Symmetric trace-free
TIGER	Test Infrastructure for GEneral Relativity
TT	Transverse-traceless

Part I
Introduction to Gravitational-Wave Data Analysis

Chapter 1
Gravitational Waves in the Linearised Theory of General Relativity

1.1 Brief Overview of General Relativity

The theory of general relativity (GR) states that gravity can be considered as curvature of spacetime [1]. This relationship is made explicit by the Einstein field equations (EFE),

$$G^{\mu\nu} \equiv R^{\mu\nu} - \frac{1}{2}g^{\mu\nu}R = 8\pi T^{\mu\nu}, \tag{1.1}$$

which form a set of ten coupled differential equations. On the left hand side (LHS), we have defined the Einstein tensor $G_{\mu\nu}$, which only depends on terms describing the geometry of spacetime. On the right hand side (RHS), one has the stress-energy tensor, which describes the density and flux of energy and momentum. From the second Bianchi identity,

$$R_{\alpha\beta\mu\nu;\lambda} + R_{\alpha\beta\lambda\mu;\nu} + R_{\alpha\beta\nu\lambda;\mu} = 0, \tag{1.2}$$

it follows that

$$G^{\mu\nu}{}_{;\mu} = 0. \tag{1.3}$$

Consequently, the stress-energy tensor satisfies

$$T^{\mu\nu}{}_{;\mu} = 0, \tag{1.4}$$

so that the energy and momentum are covariantly conserved quantities in GR.

T.G.F. Li, *Extracting Physics from Gravitational Waves*,
Springer Theses, DOI 10.1007/978-3-319-19273-4_1

1.1.1 Geodesics

As a generalisation of Newton's first law, i.e. objects move on a straight line in absence of external forces, GR states that objects move along *geodesics* in absence of external forces. Geodesics are the curved-space equivalents of straight lines, which can be found by parallel transporting the tangent vector of a curve. Therefore, the equations of motion within GR are given by the *geodesic equation*

$$\begin{aligned} \frac{d^2x^\mu}{d\tau^2} + \Gamma^\mu{}_{\nu\rho}\frac{dx^\nu}{d\tau}\frac{dx^\rho}{d\tau} &= 0 \quad \text{for } m \neq 0, \\ \frac{d^2x^\mu}{d\lambda^2} + \Gamma^\mu{}_{\nu\rho}\frac{dx^\nu}{d\lambda}\frac{dx^\rho}{d\lambda} &= 0 \quad \text{for } m = 0. \end{aligned} \tag{1.5}$$

where m is the mass of the object, τ represents the proper time given by $d\tau^2 = -ds^2$, and λ is some affine parameter on the geodesic.

In a flat spacetime, two straight lines that are initially parallel to each other will remain parallel. In a curved spacetime, geodesics do not satisfy this property. Instead, two nearby geodesics, separated by ζ^μ, follow the *geodesic deviation equation*

$$\frac{D^2\zeta^\mu}{D\tau^2} = -R^\mu{}_{\nu\rho\sigma}\zeta^\rho\frac{dx^\nu}{d\tau}\frac{dx^\sigma}{d\tau}, \tag{1.6}$$

where $D/D\tau$ is defined as

$$\frac{DV^\mu}{D\tau} \equiv \frac{dV^\mu}{d\tau} + \Gamma^\mu{}_{\nu\rho}V^\nu\frac{dx^\rho}{d\tau}, \tag{1.7}$$

and denotes the covariant derivative along a curve that is parameterised by τ. The geodesic deviation equation describes the change in separation ζ^μ between two nearby geodesics. As the Riemann tensor describes the tidal forces caused by a gravitational field, Eq. (1.6) shows that these tidal forces can be considered as deviations of nearby geodesics.

1.2 Linearised Theory of General Relativity

To study the properties of gravitational waves (GWs), it is instructive to first study them in situations where the gravitational fields are weak. In the so-called *weak-field approximation*, one can view the metric as the Minkowski metric with a small perturbation. In this chapter, we will discuss the properties of GWs in the linearised theory of GR, which only considers the lowest order perturbation. In particular, we will look into the propagation of GWs on the flat background spacetime, their interaction with test masses, the energy and momentum carried by GWs, and the generation of gravitational waves GWs.

1.2.1 Weak-Field Metric

In GR, the EFE are covariant under general coordinate transformations

$$x^{\mu} \rightarrow x^{\mu'}(x), \tag{1.8}$$

so that the metric transforms as

$$g_{\mu\nu} \rightarrow g_{\mu'\nu'} = x^{\rho}_{,\mu'} x^{\sigma}_{,\nu'} g_{\rho\sigma}. \tag{1.9}$$

This means that one is free to choose a convenient coordinate system without altering the physical predictions of the EFE. In the weak-field regime, where spacetime is nearly flat, it is convenient to choose the coordinates such that the metric is composed of the flat spacetime metric with a small perturbation, i.e.

$$g_{\mu\nu} = \eta_{\mu\nu} + h_{\mu\nu}, \quad \text{for } \left|h_{\mu\nu}\right| \ll 1. \tag{1.10}$$

However, by choosing the coordinates such that Eq. (1.10) holds, the general coordinate invariance of general relativity GR is broken. Instead, one has a smaller set of transformations that does preserve Eq. (1.10). We will discuss these in turn.

Gauge Transformations

Firstly, consider gauge transformations of the form

$$x^{\mu} \rightarrow x^{\mu'} = \eta^{\mu'}{}_{\nu}(x^{\nu} + \xi^{\nu}) \tag{1.11}$$

where we demand that

$$\left|\xi_{\mu,\nu}\right| \lesssim \left|h_{\mu\nu}\right|. \tag{1.12}$$

Consequently, the metric in Eq. (1.10) transforms as

$$\begin{aligned} g_{\mu\nu} \rightarrow g_{\mu'\nu'} &= x^{\rho}_{,\mu'} x^{\sigma}_{,\nu'} g_{\rho\sigma} \\ &= \eta_{\mu\nu} + h_{\mu\nu} - \xi_{\mu,\nu} - \xi_{\nu,\mu} + \cdots, \end{aligned} \tag{1.13}$$

where $\eta_{\mu\nu}$ lowers and $\eta^{\mu\nu}$ raises indices in the linearised theory. Keeping the terms linear in $h_{\mu\nu}$ or $\xi_{\mu,\nu}$, we observe that $h_{\mu\nu}$ transforms as

$$h_{\mu\nu} \rightarrow h_{\mu'\nu'} = h_{\mu\nu} - \xi_{\mu,\nu} - \xi_{\nu,\mu}. \tag{1.14}$$

Therefore, provided Eq. (1.12) holds, the form of the metric as shown in Eq. (1.10) is preserved.

Lorentz Transformations
Secondly, consider Lorentz transformations of the form

$$x^{\mu} \to x^{\mu'} = \Lambda^{\mu'}{}_{\nu} x^{\nu}, \tag{1.15}$$

where $\Lambda^{\mu'}{}_{\nu}$ is the Lorentz transformation from special relativity that transforms the flat spacetime metric as

$$\Lambda^{\mu}_{\rho'} \Lambda^{\nu}_{\sigma'} \eta_{\mu\nu} = \eta_{\rho'\sigma'}. \tag{1.16}$$

The full metric then transforms as

$$\begin{aligned} g_{\mu\nu} \to g_{\mu'\nu'} &= \Lambda^{\rho}{}_{\mu'} \Lambda^{\sigma}{}_{\nu'} g_{\rho\sigma} \\ &= \eta_{\mu'\nu'} + \Lambda^{\rho}{}_{\mu'} \Lambda^{\sigma}{}_{\nu'} h_{\rho\sigma}, \end{aligned} \tag{1.17}$$

and therefore

$$h_{\mu\nu} \to h_{\mu'\nu'} = \Lambda^{\rho}{}_{\mu'} \Lambda^{\sigma}{}_{\nu'} h_{\rho\sigma}. \tag{1.18}$$

Note that $h_{\mu\nu}$ is a pseudotensor, as it only behaves as a tensor under a Lorentz transformation. However, since a boost can spoil the condition $|h_{\mu\nu}| \ll 1$, one is limited to boosts that do not violate this condition.

Similar to the use of the general coordinate freedom to simplify the EFE, we will use both the freedom to perform a gauge transformation shown in Eq. (1.11) and a Lorentz transformation shown in Eq. (1.15) to simplify equations in the linearised theory of GR.

1.2.2 Linearising the Einstein Field Equations

Next, our aim is to derive the Einstein field equations EFE in the linearised theory. To leading order in $h_{\mu\nu}$, the Christoffel symbol, Riemann tensor, Ricci tensor and Ricci scalar are given by

$$\Gamma^{\mu}{}_{\alpha\beta} = \frac{1}{2}\left(h_{\alpha}{}^{\mu}{}_{,\beta} + h_{\beta}{}^{\mu}{}_{,\alpha} - h_{\alpha\beta}{}^{,\mu}\right), \tag{1.19}$$

$$R_{\mu\nu\alpha\beta} = \frac{1}{2}\left(h_{\mu\beta,\nu\alpha} + h_{\nu\alpha,\mu\beta} - h_{\nu\beta,\mu\alpha} - h_{\mu\alpha,\nu\beta}\right), \tag{1.20}$$

$$R_{\mu\nu} = \frac{1}{2}\left(h_{\mu}{}^{\alpha}{}_{,\nu\alpha} + h_{\nu}{}^{\alpha}{}_{,\mu\alpha} - h_{\mu\nu,\alpha}{}^{\alpha} - h_{\alpha}{}^{\alpha}{}_{,\mu\nu}\right), \tag{1.21}$$

$$R = \frac{1}{2}\left(h_{\alpha\beta}{}^{,\alpha\beta} - h_{\alpha}{}^{\alpha}{}_{,\beta}{}^{\beta}\right). \tag{1.22}$$

From Eq. (1.20) it follows that the Riemann tensor is invariant under the gauge transformation of Eq. (1.11). Next, EFE can be written as

$$\begin{aligned} 8\pi T_{\mu\nu} &= R_{\mu\nu} - \frac{1}{2} R g_{\mu\nu} \\ &= \frac{1}{2}\left[h_{\mu\alpha,\nu}{}^{\alpha} + h_{\nu\alpha,\mu}{}^{\alpha} - h_{\mu\nu,\alpha}{}^{\alpha} - h_{\alpha}{}^{\alpha}{}_{,\mu\nu} - \eta_{\mu\nu}\left(h_{\alpha\beta}{}^{,\alpha\beta} - h_{\alpha}{}^{\alpha}{}_{,\beta}{}^{\beta} \right)\right]. \end{aligned} \tag{1.23}$$

One can write Eq. (1.23) in a more compact form by introducing the trace h given by

$$h \equiv \eta^{\alpha\beta} h_{\alpha\beta}, \tag{1.24}$$

and the trace-reversed $\bar{h}_{\mu\nu}$ given by

$$\bar{h}_{\mu\nu} \equiv h_{\mu\nu} - \frac{1}{2}\eta_{\mu\nu} h, \tag{1.25}$$

so that Eq. (1.23) becomes

$$16\pi T_{\mu\nu} = \bar{h}_{\mu\alpha}{}^{,\alpha}{}_{\nu} + \bar{h}_{\nu\alpha}{}^{,\alpha}{}_{\mu} - \bar{h}_{\mu\nu,\alpha}{}^{\alpha} - \eta_{\mu\nu}\bar{h}_{\alpha\beta}{}^{,\alpha\beta}. \tag{1.26}$$

Eq. (1.26) can be further simplified by using the gauge freedom of Eq. (1.11) to choose the *harmonic gauge*

$$\bar{h}_{\mu\nu,}{}^{\nu} = 0. \tag{1.27}$$

As we will see later, by imposing the harmonic gauge one has chosen the coordinates in such a way that for a single plane wave (or a superposition of plane waves with their wave vectors pointing in the same direction), the GW polarisations are perpendicular to the direction of propagation.

It turns out the harmonic gauge does not fix the gauge completely. To make this explicit, consider the gauge transform in Eq. (1.11) of $\bar{h}_{\mu\nu}$, i.e.

$$\bar{h}_{\mu\nu} \rightarrow \bar{h}_{\mu'\nu'} = \bar{h}_{\mu\nu} - \xi_{\mu,\nu} - \xi_{\nu,\mu} + \eta_{\mu\nu}\xi^{\rho}{}_{,\rho}. \tag{1.28}$$

Applying the harmonic gauge condition to Eq. (1.28), one has

$$\bar{h}_{\mu'\nu',}{}^{\nu'} = \bar{h}_{\mu\nu,}{}^{\nu} - \xi_{\mu,\nu}{}^{\nu}. \tag{1.29}$$

Therefore, if the harmonic gauge holds in the old frame, the condition

$$\xi_{\mu,\nu}{}^{\nu} = 0 \tag{1.30}$$

does not spoil the harmonic gauge and allows an additional gauge specification. We will use this property when we look for a specific form of the plane wave representation of GWs.

Finally, in the harmonic gauge, the EFE can be written in the simple form

$$-\bar{h}_{\mu\nu,\alpha}{}^{\alpha} = 16\pi T_{\mu\nu}. \tag{1.31}$$

This equation forms the backbone of the linearised theory of GR and we will use it to study some of the properties of GWs.

How many degrees of freedom do we expect to be present in the linearised EFE? The symmetric metric tensor $g_{\mu\nu}$ has ten independent components. By imposing the harmonic gauge, we have introduced four constraints. Further specifying the gauge by using gauge transformations that satisfy Eq. (1.30) introduces an additional four constraints, leaving the metric tensor with two independent degrees of freedom. Indeed, these eight constraints are in line with the set of transformations that preserve the form of the metric in Eq. (1.10), namely the gauge transformation (four constraints) and the Lorentz transformation (four constraints).

1.3 Propagation of Gravitational Waves in Vacuum

To gain insight into the propagation of GWs in the linearised theory, we consider the linearised EFE in vacuo,

$$\bar{h}_{\mu\nu,\alpha}{}^{\alpha} = 0. \tag{1.32}$$

The solution to Eq. (1.32) can be written in the form of plane waves

$$\bar{h}_{\mu\nu} = \Re\left[A_{\mu\nu}e^{\imath k_\alpha x^\alpha}\right], \tag{1.33}$$

where $\Re$ denotes the real part of a quantity. The linearised Ei EFE then yield

$$k_\alpha k^\alpha = 0, \tag{1.34}$$

so that the wave vector k_α is a null vector and the wave travels on the light cone of the flat background metric. Furthermore, the harmonic gauge in Eq. (1.27) also imposes the condition

$$A_{\mu\nu}k^\nu = 0, \tag{1.35}$$

showing that the direction propagation, k_α, is orthogonal to the wave polarization, $A_{\mu\nu}$. In other words, the wave is *transverse* to the direction of propagation.

1.3.1 Transverse-Traceless Gauge

One can now use the residual gauge freedom within the harmonic gauge class, as shown in Eq. (1.30), to further specify $A_{\mu\nu}$. Suppose one specifies a fixed 4-velocity u^μ throughout all of spacetime (from a special-relativity perspective). In other words, one defines a family of observers, one on each point in spacetime, that all have a 4-velocity u^μ. A choice of coordinates is then made by imposing

$$A_{\mu\nu}u^\nu = 0. \tag{1.36}$$

This condition gives three constraints as one of them can be identified with $k^\mu \left(A_{\mu\nu}u^\nu\right) = 0$. As the fourth constraint, one can use the residual gauge freedom to set

$$A_\mu{}^\mu = 0. \tag{1.37}$$

One can now use a Lorentz transformation to make the explicit choice that the conditions in Eqs. (1.36) and (1.37) hold in a frame where $u^0 = 1$ and $u^j = 0$. With this choice, one has specified all the free components. The metric perturbation satisfies

$$\begin{aligned} \bar{h}_{\mu 0} &= 0 \quad \text{(only spatial components)}, \\ \bar{h}_j{}^j &= 0 \quad \text{(traceless in the spatial part)}, \\ \bar{h}_{ij,}{}^j &= 0 \quad \text{(spatial part is divergence free)}. \end{aligned} \tag{1.38}$$

The conditions in Eq. (1.38) define the transverse-traceless (TT) gauge. In the TT gauge, one has $h = -\bar{h} = -\bar{h}_\mu{}^\mu = -\bar{h}_j{}^j = 0$, so that $h_{\mu\nu}$ is the same as $\bar{h}_{\mu\nu}$. Finally, quantities in this gauge are denoted by the superscript TT, e.g. $h_{\mu\nu}^{\mathrm{TT}}$.

Next, consider a single plane wave with wave vector $k^\mu = (\omega, k^i)$, where $\omega = \left|k^i\right|$. Without loss of generality, one can set the direction of propagation to be the z-direction. Under these conditions, $h_{\mu\nu}^{\mathrm{TT}}$ can be written as

$$h_{\mu\nu}^{\mathrm{TT}} = \begin{pmatrix} 0 & 0 & 0 & 0 \\ 0 & h_+ & h_\times & 0 \\ 0 & h_\times & -h_+ & 0 \\ 0 & 0 & 0 & 0 \end{pmatrix} \cos\left[\omega(t - z)\right], \tag{1.39}$$

where h_+ and $h_\times$ are the two degrees of freedom in the linearised theory of GR, corresponding to two polarisations of a GW. This nomenclature will be clarified in Sect. 1.4. The line element in the TT frame is given by

$$\begin{aligned} ds^2 = &- dt^2 + dz^2 + \{1 + h_+ \cos\left[\omega(t - z)\right]\} dx^2 + 2h_\times \cos\left[\omega(t - z)\right] dxdy \\ &+ \{1 - h_+ \cos\left[\omega(t - z)\right]\} dy^2. \end{aligned} \tag{1.40}$$

1.4 Interaction with Test Masses

Next, we study the effect of GWs on test masses (mirrors in the case of the interferometric detectors, see Fig. 3.1), by using the geodesic equation and the geodesic deviation equation, shown in Eqs. (1.5) and (1.6) respectively.

1.4.1 Transverse-Traceless Gauge

Consider a test mass initially at rest at $\tau = 0$. From Eq. (1.5) it follows that

$$\begin{aligned} \frac{d^2x^i}{d\tau^2} &= -\left[\Gamma^i{}_{\nu\rho}\frac{dx^\nu}{d\tau}\frac{dx^\rho}{d\tau}\right]_{\tau=0} \\ &= -\left[\Gamma^i{}_{00}\left(\frac{dx^0}{d\tau}\right)^2\right]_{\tau=0}. \end{aligned} \tag{1.41}$$

Evaluating Eq. (1.19) in the TT gauge, one has

$$\begin{aligned} \Gamma^i{}_{00} &= \frac{1}{2}\left(2h_0{}^i{}_{,0} - h_{00,}{}^i\right) \\ &= 0, \end{aligned} \tag{1.42}$$

where the TT superscript is dropped for convenience. Hence, a test mass initially at rest will remain at rest as viewed in the TT frame. Similarly, consider a coordinate separation ζ^i between two test masses that are initially at rest with respect to each other at $\tau = 0$. The geodesic deviation equation in Eq. (1.6) gives us

$$\begin{aligned} \left.\frac{d^2\zeta^i}{d\tau^2}\right|_{\tau=0} &= -\left[2\Gamma^i{}_{0j}\frac{d\zeta^j}{d\tau}\right]_{\tau=0} \\ &= 0. \end{aligned} \tag{1.43}$$

This shows that if two test masses are initially separated by a coordinate separation of ζ^i and are at rest with respect to each other, they will remain at this separation. Overall, it seems that a GW has no influence on the geodesic or on the deviation of geodesics. As we will see next, this is merely a feature of the TT coordinate system, as we have explicitly chosen $u^0 = 1$ and $u^j = 0$ to hold throughout spacetime.

Consider two events in the TT frame at $(t, x_1, 0, 0)$ and $(t, x_2, 0, 0)$. These two events are separated by the coordinate distance $x_2 - x_1 = L$, where L is a constant with respect to the coordinate time t. However, physical effects are encoded in the *proper distance*, given by

$$
\begin{aligned}
s &= \int ds \\
&= \int_{x_1}^{x_2} dx \, \sqrt{1 + h_+ \cos \omega t} \\
&\approx L \left(1 + \frac{1}{2} h_+ \cos \omega t \right) .
\end{aligned} \tag{1.44}
$$

In general directions, the proper distance is given by

$$
s = \sqrt{L^2 + h_{ij}(t) L_i L_j}, \tag{1.45}
$$

where L^i denotes the spatial separation between two test masses and L the associated coordinate distance. Therefore, the proper distance expands and shrinks periodically. Because the time that light travels between the two test masses is related to the proper distance, GWs leave an imprint on the time it takes for a photon to make a round trip. Consequently, interferometers can potentially measure these imprints by measuring the length difference between their arms (see Chap. 3 for more information about detecting GWs by using interferometers).

1.4.2 Proper Detector Frame

Alternatively, consider a detector capable of measuring changes in the proper distance, e.g. an interferometer, with a characteristic size that is much smaller than the characteristic wavelength of the GW. In this case, one can approximate the entire detector to be in a near local Lorentz frame LLF (freely falling frame), even in the presence of GWs. The metric in the near local Lorentz frame LLF can be written as

$$
ds^2 \approx -dt^2 + \delta_{ij} dx^i dx^j + \mathcal{O}\left(\frac{x^i x^j}{L_B^2} \right), \tag{1.46}
$$

where L_B denotes the typical variation scale of the metric. Consider two test masses in free fall separated by ζ^i. We want to know the influence of GWs on these two test masses. It is instructive to then rewrite the geodesic deviation equation given by Eq. (1.6) as

$$
\frac{d^2 \zeta^\mu}{d\tau^2} + 2\Gamma^\mu{}_{\nu\rho} \frac{dx^\nu}{d\tau} \frac{dx^\rho}{d\tau} + \zeta^\sigma \Gamma^\mu{}_{\nu\rho,\sigma} \frac{dx^\nu}{d\tau} \frac{dx^\rho}{d\tau} = 0. \tag{1.47}
$$

Next, assume that the two test masses are moving non-relativistically so that $dx^i/d\tau$ can be neglected compared to $dx^0/d\tau$. Furthermore, the term proportional to $\Gamma^\mu{}_{\nu\rho}$ is negligible compared to the other terms in a near LLF. Consequently, one has

$$\frac{d^2\zeta^i}{d\tau^2} + \zeta^\sigma {\Gamma^i}_{00,\sigma} \left(\frac{dx^0}{d\tau}\right)^2 = 0. \tag{1.48}$$

One can further simplify $\zeta^\sigma {\Gamma^i}_{00,\sigma} \approx \zeta^j {\Gamma^i}_{00,j}$ and thus write

$$\frac{d^2\zeta^i}{d\tau^2} + \zeta^j {\Gamma^i}_{00,j} \left(\frac{dx^0}{d\tau}\right)^2 = 0. \tag{1.49}$$

But in the LLF, ${R^i}_{0j0} = {\Gamma^i}_{00,j} - {\Gamma^i}_{0j,0} = {\Gamma^i}_{00,j}$ and therefore

$$\frac{d^2\zeta^i}{d\tau^2} + {R^i}_{0j0}\zeta^j \left(\frac{dx^0}{d\tau}\right)^2 = 0. \tag{1.50}$$

Because $dx^0/d\tau \approx 1$, one can approximate $\tau \approx t$. This allows us to write

$$\ddot{\zeta}^j = -{R^i}_{0j0}\zeta^j, \tag{1.51}$$

where an overdot denotes a derivative with respect to the coordinate time t. Finally, instead of evaluating the Riemann tensor in this frame, we use the fact that the Riemann tensor is invariant under the gauge transformation in Eq. (1.11). Therefore, we can evaluate Eq. (1.20) in the TT gauge and write

$${R^i}_{0j0} = R_{i0j0} = -\frac{1}{2}\ddot{h}_{ij}^{\mathrm{TT}}. \tag{1.52}$$

Inserting Eq. (1.52) into Eq. (1.51), the geodesic deviation equation in the proper detector frame takes the form

$$\ddot{\zeta}^i = \frac{1}{2}\ddot{h}_{ij}^{\mathrm{TT}}\zeta^j. \tag{1.53}$$

Remarkably, the influence of a GW in a near LLF resembles a *Newtonian force*, allowing us to think about it in terms of classical mechanics.

Caution must be taken with the use of the geodesic deviation equation and the result in Eq. (1.53) that follows from it. The geodesic deviation equation in Eq. (1.6) is only valid if ζ is small compared to the typical variation scale of the metric. When the variation of the metric is caused by GWs, the scale at which the metric varies is the reduced wavelength of the GW, $\lambdabar = \lambda/(2\pi)$. Therefore, the result in Eq. (1.53) is only valid if

$$L \ll \lambdabar, \tag{1.54}$$

where L denotes the typical length scale of the GW detector. For LIGO or Virgo ($\lambdabar \approx 10^5\,\mathrm{m}$ and $L \approx 10^3\,\mathrm{m}$), the geodesic deviation equation is valid. For LISA

($\lambda \approx 10^{10}$ m and $L \approx 10^9$ m), the geodesic deviation equation is no longer valid and a full GR treatment is needed to study the influence of GWs on test masses.

1.4.3 Ring of Test Masses

Consider a ring of test masses in the (x, y) plane centred at $z = 0$ and a GW travelling in the z-direction. Because h_{ij}^{TT} is transverse to the propagation direction, the GW will only have influence in the plane of the test masses and we can restrict our attention to the (x, y) plane alone. Furthermore, as the two polarisations of Eq. (1.39) are independent, one can consider them separately. For example, the plus polarisation is given by

$$h_{ab}^{\mathrm{TT}} = h_+ \cos \omega t \begin{pmatrix} 1 & 0 \\ 0 & -1 \end{pmatrix}. \tag{1.55}$$

Consider a proper detector frame where the location of a test mass is denoted by $\zeta^a(t) = (x_0 + \delta x(t), y_0 + \delta y(t))$, and where (x_0, y_0) is the location of the unperturbed test mass and $(\delta x, \delta y)$ is the perturbation due to the GW. The evolution of $\zeta^a(t)$ caused by the plus polarisation is then given by Eq. (1.53)

$$\begin{aligned} \begin{pmatrix} \ddot{\delta x} \\ \ddot{\delta y} \end{pmatrix}_+ &= -\frac{h_+}{2} \begin{pmatrix} 1 & 0 \\ 0 & -1 \end{pmatrix} \begin{pmatrix} x_0 + \delta x \\ y_0 + \delta y \end{pmatrix} \omega^2 \cos \omega t \\ &\approx -\frac{h_+}{2} \begin{pmatrix} x_0 \\ -y_0 \end{pmatrix} \omega^2 \cos \omega t, \end{aligned} \tag{1.56}$$

where one assumes that the perturbations are $\mathcal{O}(h)$, and thus small compared to the unperturbed locations. Solving Eq. (1.56), the deviations caused by the plus polarisations are given by

$$\begin{pmatrix} \delta x \\ \delta y \end{pmatrix}_+ = \frac{h_+}{2} \begin{pmatrix} x_0 \\ -y_0 \end{pmatrix} \cos \omega t. \tag{1.57}$$

Similarly, the influence of the cross polarisation on a ring of test masses can be described by

$$\begin{pmatrix} \delta x \\ \delta y \end{pmatrix}_\times = \frac{h_\times}{2} \begin{pmatrix} y_0 \\ x_0 \end{pmatrix} \cos \omega t. \tag{1.58}$$

These deformations can be visualised by considering a ring of test masses as shown in Fig. 1.1. The labels "plus" and "cross" originate from the directions of the motion of such a ring of test masses.

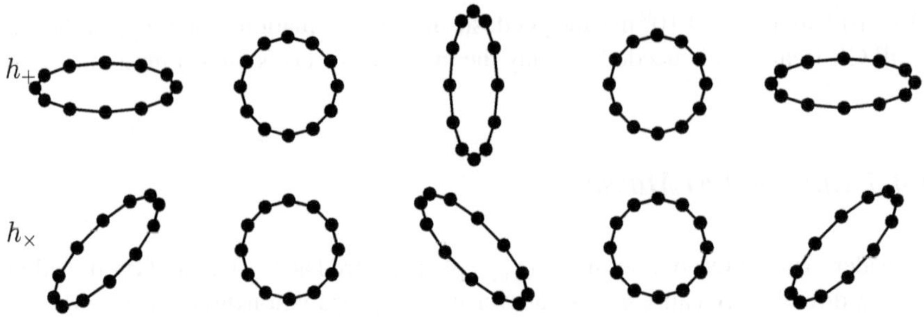

Fig. 1.1 Deformations of a ring of test masses caused by the plus (+) or cross (×) polarisations of a GW for various values of the phase

1.5 Energy and Momentum of Gravitational Waves

From the motion of test masses, one has a clue that GWs carry energy and momentum. However, the formalism in Sects. 1.2–1.4 only describes the propagation of GWs on a flat spacetime, and does not describe the energy and momentum associated to GWs. Indeed, enforcing a flat background metric, as shown in Eq. (1.10), does not allow GWs to curve the background metric.

1.5.1 Non-linear Corrections

To study the energy and momentum carried by GWs, one must extend the approximation from Eq. (1.10) to

$$g_{\mu\nu} = g^{(\mathrm{B})}_{\mu\nu} + h^{(1)}_{\mu\nu} + h^{(2)}_{\mu\nu}, \tag{1.59}$$

where $g^{(\mathrm{B})}_{\mu\nu}$ is the background metric that is now allowed to be curved, and $h^{(1)}_{\mu\nu}$ and $h^{(2)}_{\mu\nu}$ are the first and second order perturbations describing the GW. Within this approximation, the Ricci tensor can be written as

$$R_{\mu\nu} = R^{(\mathrm{B})}_{\mu\nu} + R^{(1)}_{\mu\nu}\left[h^{(1)}_{\mu\nu}\right] + R^{(2)}_{\mu\nu}\left[h^{(1)}_{\mu\nu}\right] + R^{(1)}_{\mu\nu}\left[h^{(2)}_{\mu\nu}\right], \tag{1.60}$$

where $R^{(\mathrm{B})}_{\mu\nu}$ is the Ricci tensor for the background metric $g^{(\mathrm{B})}_{\mu\nu}$. The linear and quadratic corrections to the Ricci tensor are given by $R^{(1)}_{\mu\nu}$ and $R^{(2)}_{\mu\nu}$ respectively. For example, $R^{(2)}_{\mu\nu}\left[h^{(1)}_{\mu\nu}\right]$ represents those terms in the Ricci tensor that are of second order with respect to the first order perturbation of the background metric, given by $h^{(1)}_{\mu\nu}$.

1.5.2 Short-Wave Approximation

Although it is valid to approximate the metric as Eq. (1.59), one runs into the problem of distinguishing the perturbation from the background. In fact, there is no generally unambiguous way of distinguishing a perturbation from the background. Consequently, the notion of local energy density associated to GWs does not exist.

Nevertheless, a distinction in terms of the characteristic lengths can be made for detection of GWs on Earth. We introduce two length scales: $\lambda\!\!\!^{-}$ denotes the scale on which $h_{\mu\nu}$ varies, whereas $\mathcal{R}$ denotes the scale on which $g^{(B)}_{\mu\nu}$ varies. We assume that the GWs measured on Earth have much smaller characteristic length scales than the background metric, i.e. $\lambda\!\!\!^{-} \ll \mathcal{R}$. This is often called the *short-wave approximation*. Within the short-wave approximation, one can introduce a scheme to separate the background (smooth) from the perturbation (fluctuating). Such a split can be accomplished by introducing a length scale l such that $\lambda\!\!\!^{-} \ll l \ll \mathcal{R}$.

We can now introduce the average over a spatial volume with sides of length l, denoted by $\langle \cdots \rangle_l$. The effect of such an averaging scheme is that terms that vary on the scale of $\lambda\!\!\!^{-}$ average to zero, whereas terms that vary on the scale of $\mathcal{R}$ remain constant. For example, the part of $R^{(2)}_{\mu\nu}$ that varies on the scale of $\mathscr{R}$ can be written as $\left\langle R^{(2)}_{\mu\nu} \right\rangle_l$. On the other hand, the part that fluctuates on the scale of $\lambda\!\!\!^{-}$ can be identified as the remainder, $R^{(2)}_{\mu\nu} - \left\langle R^{(2)}_{\mu\nu} \right\rangle_l$. Similarly, the Ricci tensor can thus be split into two parts: a smooth (associated to $\mathscr{R}$) and a fluctuating part (associated to $\lambda\!\!\!^{-}$). Up to second order contributions in $h_{\mu\nu}$, the Ricci tensor can be split into

$$R^{(\text{smooth})}_{\mu\nu} = R^{(B)}_{\mu\nu} + \left\langle R^{(2)}_{\mu\nu}\left[h^{(1)}_{\mu\nu}\right]\right\rangle_l , \tag{1.61}$$

$$R^{(\text{fluc})}_{\mu\nu} = R^{(1)}_{\mu\nu}\left[h^{(1)}_{\mu\nu} + h^{(2)}_{\mu\nu}\right] + R^{(2)}_{\mu\nu}\left[h^{(1)}_{\mu\nu}\right] - \left\langle R^{(2)}_{\mu\nu}\left[h^{(1)}_{\mu\nu}\right]\right\rangle_l . \tag{1.62}$$

1.5.3 Stress-Energy Tensor of Gravitational Waves

Next, we want to investigate how GWs impact the background metric. To make such a relationship explicit, consider the smooth part of the Einstein field equations EFE in vacuo by only using the smooth part of the Ricci tensor given by Eq. (1.61). One can rewrite the vacuum EFE as

$$8\pi T^{(\text{GW})}_{\mu\nu} = R^{(B)}_{\mu\nu} - \frac{1}{2} R^{(B)} g^{(B)}_{\mu\nu} , \tag{1.63}$$

where the stress-energy tensor of GWs is defined as

$$T^{(\text{GW})}_{\mu\nu} \equiv -\frac{1}{8\pi}\left\langle R^{(2)}_{\mu\nu} - \frac{1}{2} g^{(B)}_{\mu\nu} R^{(2)} \right\rangle_l , \tag{1.64}$$

and the quadratic order correction to the Ricci tensor is given by

$$R^{(2)}_{\mu\nu}\left[h_{\mu\nu}\right] \equiv \frac{1}{2}\left[\frac{1}{2}h_{\alpha\beta;\mu}h^{\alpha\beta}{}_{;\nu} + h_{\nu}{}^{\alpha;\beta}\left(h_{\alpha\mu;\beta} - h_{\beta\mu;\alpha}\right)\right.$$
$$+ h^{\alpha\beta}\left(h_{\alpha\beta;\mu\nu} + h_{\mu\nu;\alpha\beta} - h_{\alpha\mu;\nu\beta} - h_{\alpha\nu;\mu\beta}\right)$$
$$\left. - \left(h^{\alpha\beta}{}_{;\beta} - \frac{1}{2}h^{;\alpha}\right)\left(h_{\alpha\mu;\nu} + h_{\alpha\nu;\mu} - h_{\mu\nu;\alpha}\right)\right]. \tag{1.65}$$

One can thus see that the energy associated to GWs is only defined as an average over several periods of the characteristic wavelength. One can further simplify Eq. (1.64) by noting/assuming the following:

- Assume that the background is close to flat so that one can write $A_{;\mu} \approx A_{,\mu}$, where A can be a vector or a tensor.
- Boundary terms can be ignored in the integration by parts because $\lambda\!\!\!^{-} \ll l$.
- Time and space derivatives can be interchanged, as solutions of $\bar{h}_{\mu\nu,\alpha}{}^{\alpha} = 0$ depend on the retarded time $t - z$.
- Use the field equation $\bar{h}_{\mu\nu,\alpha}{}^{\alpha} = 0$, and the gauge conditions $\bar{h}_{\mu\nu,}{}^{\nu} = 0$ and $\bar{h} = 0$.

With these approximations, the stress-energy tensor takes the simple form

$$T^{(\mathrm{GW})}_{\mu\nu} = \frac{1}{32\pi}\left\langle \bar{h}_{\alpha\beta,\mu}\bar{h}^{\alpha\beta}{}_{,\nu}\right\rangle_l . \tag{1.66}$$

In this form, $T^{(\mathrm{GW})}_{\mu\nu}$ can be shown to be invariant under Eq. (1.11) and conserved when observed far away from the source. Under these conditions and explicitly choosing the TT gauge, the energy and momentum radiated per unit time through a sphere of radius r are given by

$$\frac{dE_{(\mathrm{GW})}}{dt} = \frac{r^2}{32\pi}\int d\Omega \left\langle \dot{h}^{\mathrm{TT}}_{ij}\dot{h}^{\mathrm{TT}}_{ij}\right\rangle_l , \tag{1.67}$$
$$\frac{dP^k_{(\mathrm{GW})}}{dt} = -\frac{r^2}{32\pi}\int d\Omega \left\langle \dot{h}^{\mathrm{TT}}_{ij}\partial^k h^{\mathrm{TT}}_{ij}\right\rangle_l . \tag{1.68}$$

1.6 Generation of Gravitational Waves

Recall that the linearised EFE in the harmonic gauge can be written as

$$\bar{h}_{\mu\nu,\alpha}{}^{\alpha} = -16\pi T_{\mu\nu}. \tag{1.69}$$

The solution to this equation can be found by using a Green's function with the appropriate boundary conditions. When considering the generation of radiation, the appropriate condition boundary is to impose that there is no incoming radiation.

The solution can then be written as a function of the retarded time and the spatial position

$$\bar{h}_{\mu\nu}(t,\vec{x}) = 4\int d^3\vec{y}\,\frac{T_{\mu\nu}(t-|\vec{x}-\vec{y}|,\vec{y})}{|\vec{x}-\vec{y}|}, \tag{1.70}$$

where $\vec{y}$ is restricted to the inside of the source. This solution can be projected onto the TT frame by introducing the projecting operator

$$\Lambda_{ijkl} \equiv P_{ik}P_{jl} - \frac{1}{2}P_{ij}P_{kl}, \tag{1.71}$$

where P_{ij} is defined as

$$P_{ij} \equiv \delta_{ij} - n_i n_j, \tag{1.72}$$

and n_i is the direction of propagation of the GW. Provided that the perturbation is in the harmonic gauge, the components in the TT gauge can be obtained by

$$\begin{aligned} h_{ij}^{\mathrm{TT}}(t,\vec{x}) &= \Lambda_{ijkl}\bar{h}_{kl} \\ &= 4\Lambda_{ijkl}\int d^3\vec{y}\,\frac{T_{kl}(t-|\vec{x}-\vec{y}|,\vec{y})}{|\vec{x}-\vec{y}|}. \end{aligned} \tag{1.73}$$

1.6.1 Multipole Expansion

One can further simplify the stress-energy tensor by imposing that the typical velocity v of the source is non-relativistic, i.e.

$$\lambda\!\!\!^{-} \sim \frac{c}{v}d \gg d, \tag{1.74}$$

where $\lambda\!\!\!^{-} = \lambda/2\pi$ and d represent the reduced wavelength of the GW and the characteristic size of the source respectively. In this low velocity regime, one can expand the source as

$$\begin{aligned} T_{kl}(t-r+\vec{y}\cdot\vec{n},\vec{y}) \approx{}& T_{kl}(t-r,\vec{y}) + y^i n^i T_{kl,0}(t-r,\vec{y}) \\ &+ \frac{1}{2}y^i y^j n^i n^j T_{kl,00}(t-r,\vec{y}) + \cdots, \end{aligned} \tag{1.75}$$

and introduce the moments of the stress tensor

$$S^{ij}(t) = \int d^3\vec{y}\; T^{ij}(t,\vec{y}), \tag{1.76}$$

$$S^{ijk}(t) = \int d^3\vec{y}\ T^{ij}(t, \vec{y})\, y^k, \tag{1.77}$$

$$S^{ijkl}(t) = \int d^3\vec{y}\ T^{ij}(t, \vec{y})\, y^k y^l. \tag{1.78}$$

Inserting these expressions into Eq. (1.73), one obtains

$$h_{ij}^{\rm TT}(t, \vec{x}) = \frac{4\Lambda_{ijkl}}{r}\left[S^{kl} + n_m \dot{S}^{klm} + \frac{1}{2} n_m n_p \ddot{S}^{klmp} + \cdots\right]_{\rm ret}. \tag{1.79}$$

where the expression inside $[\cdots]_{\rm ret}$ is evaluated at the retarded time $u = t - r$. As $y^k \sim \mathcal{O}(d)$ and the time derivative is $\mathcal{O}(\omega)$, where ω is some typical internal frequency of the source, each successive order in Eq. (1.79) has an order of v/c more compared to the previous order. In other words, one has effectively performed an expansion in orders of v/c.

Recall that in the linearised theory, energy and momentum conservation is given by $T_{\mu\nu,}{}^{\nu} = 0$. Therefore, one can use the moments of the energy density (M) and linear momentum (P)

$$\begin{aligned}
M &= \int d^3\vec{y}\ T^{00}(t, \vec{y}), & P^i &= \int d^3\vec{y}\ T^{0i}(t, \vec{y}), \\
M^i &= \int d^3\vec{y}\ T^{00}(t, \vec{y})\, y^i, & P^{ij} &= \int d^3\vec{y}\ T^{0i}(t, \vec{y})\, y^j, \\
M^{ij} &= \int d^3\vec{y}\ T^{00}(t, \vec{y})\, y^i y^j, & P^{ijk} &= \int d^3\vec{y}\ T^{0i}(t, \vec{y})\, y^j y^k,
\end{aligned} \tag{1.80}$$

to rewrite the moments of the stress tensor. For the lowest two orders, one can write

$$S^{ij} = \frac{1}{2}\ddot{M}^{ij}, \tag{1.81}$$

$$\dot{S}^{ijk} = \frac{1}{6}\dddot{M}^{ijk} + \frac{1}{3}\left(\ddot{P}^{ijk} + \ddot{P}^{jik} - 2\ddot{P}^{kij}\right). \tag{1.82}$$

1.6.2 Mass Quadrupole

We conclude this section by looking at the leading order of the multipole expansion in more detail. Starting from Eqs. (1.79) and (1.81), we write the metric perturbation as

$$\left[h_{ij}^{\rm TT}(t, \vec{x})\right]_{\rm quad} = \frac{2}{r}\Lambda_{ijkl}\ddot{M}^{kl}(u), \tag{1.83}$$

where Λ_{ijkl} is the projection operator defined in Eq. (1.71), and $u = t - r$ is the retarded time. This leading order is commonly referred to as *mass quadrupole radiation*. One can express Eq. (1.83) in terms of the *reduced quadrupole moment*

$$Q^{ij} \equiv M^{ij} - \frac{1}{3}\delta^{ij} M_{kk}, \tag{1.84}$$

so that the metric perturbation can be written as

$$\begin{aligned} \left[h_{ij}^{\mathrm{TT}}(t, \vec{x})\right]_{\mathrm{quad}} &= \frac{2}{r}\Lambda_{ijkl}\ddot{Q}^{kl}(u), \\ &\equiv \frac{2}{r}\ddot{Q}_{ij}^{\mathrm{TT}}(u), \end{aligned} \tag{1.85}$$

where we have made use of the fact that $\Lambda_{ijkl}\ddot{M}^{kl} = \Lambda_{ijkl}\ddot{Q}^{kl}$, as the second term in Eq. (1.84) vanishes under the Λ_{ijkl} operator.

One can now construct expressions for h_+ and $h_\times$. First, we set up a Cartesian coordinate system such that n_i coincides with the z-axis. Evaluating Eq. (1.85) and comparing this to Eq. (1.39), we can conclude that

$$h_+ = \frac{\ddot{M}_{11} - \ddot{M}_{22}}{r}, \tag{1.86}$$

$$h_\times = \frac{2\ddot{M}_{12}}{r}. \tag{1.87}$$

Finally, the energy and momentum radiated can be found by inserting Eq. (1.85) into Eqs. (1.67) and (1.68)

$$\frac{dE}{dt} = \frac{1}{8\pi}\int d\Omega\, \dddot{Q}_{ij}^{\mathrm{TT}}\dddot{Q}_{ij}^{\mathrm{TT}}, \tag{1.88}$$

$$\frac{dP^k}{dt} = -\frac{1}{8\pi}\int d\Omega\, \dddot{Q}_{ij}^{\mathrm{TT}}\partial^k \ddot{Q}_{ij}^{\mathrm{TT}}. \tag{1.89}$$

Quadrupole Radiation from Binary Systems

Finally, we will look at the leading order GW emission by binary systems in the linearised theory. Binary systems are the focus of this book and are considered one of the prime candidates for the first direct detection of GWs (see Chap. 3).

Consider two masses, m_1 and m_2, in a circular orbit with an angular velocity of ω around their common centre of mass. The distance between the two objects and the distance between the binary and the observer are denoted by R and r respectively. We set up a Cartesian coordinate system in such a way that the observer is in the direction of the z-axis, the orbital plane intersects the x-axis and the origin coincides with the centre of mass. The angle between the normal to the plane of the two masses and the observer is denoted by ι. A schematic overview is shown in Fig. 1.2.

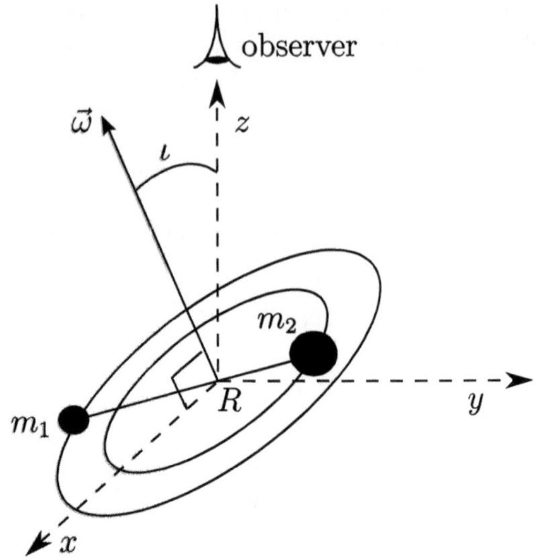

Fig. 1.2 Schematic overview of a binary system with component masses m_1 and m_2, separation between the masses R, and an orbital frequency ω. The direction of the observer coincides with the z-axis. The angle between the normal of the binary plane and the observer is denoted by ι

In this coordinate system, the location of the masses are given by

$$\vec{y}_1(t) = \frac{\mu}{m_1} R\hat{e}(t), \tag{1.90}$$

$$\vec{y}_2(t) = -\frac{\mu}{m_2} R\hat{e}(t), \tag{1.91}$$

where $\mu = m_1 m_2/(m_1 + m_2)$ and $\hat{e}(t) = (\cos(\omega t), \cos(\iota)\sin(\omega t), \sin(\iota)\sin(\omega t))$. For simplicity, we will only consider the radiation from the mass quadrupole

$$\begin{aligned} M^{ij}(t) &= \int d^3\vec{y}\, T^{00}(t, \vec{y})\, y^i y^j \\ &\approx \int d^3\vec{y}\, \left[m_1 \delta(\vec{y} - \vec{y}_1) + m_2 \delta(\vec{y} - \vec{y}_2)\right] y^i y^j \\ &\approx \mu R^2 \begin{pmatrix} \cos^2 \omega t & \cos\iota \cos\omega t \sin\omega t \\ \cos\iota\cos\omega t \sin\omega t & \cos^2\iota \sin^2\omega t \end{pmatrix}. \end{aligned} \tag{1.92}$$

Inserting this into Eqs. (1.86) and (1.87), one obtains

$$h_+ = \frac{4\mu R^2\omega^2}{r} \frac{1 + \cos^2\iota}{2} \cos(2\omega t_{\mathrm{ret}}), \tag{1.93}$$

$$h_\times = \frac{4\mu R^2\omega^2}{r} \cos\iota \sin(2\omega t_{\mathrm{ret}}). \tag{1.94}$$

Finally, we assume that the two masses are sufficiently far apart and use the Newtonian centripetal force to write R in terms of ω, m_1 and m_2. For m_1, one has

$$\frac{m_1(R\mu/m_1)^2\omega^2}{R\mu/m_1} = \frac{m_1 m_2}{R^2}, \tag{1.95}$$

and a similar expression for m_2 exists. Solving for R and defining the so-called *chirp mass*

$$\mathcal{M}_c = \frac{(m_1 m_2)^{3/5}}{(m_1 + m_2)^{1/5}}, \tag{1.96}$$

one can rewrite Eqs. (1.93) and (1.94) as

$$h_+ = \frac{4\mathcal{M}_c^{5/3}\omega^{2/3}}{r}\frac{1+\cos^2\iota}{2}\cos(2\omega t), \tag{1.97}$$

$$h_\times = \frac{4\mathcal{M}_c^{5/3}\omega^{2/3}}{r}\cos\iota\sin(2\omega t). \tag{1.98}$$

We can see that the radiation is monochromatic with a constant amplitude. Furthermore, the source radiates at *twice* the orbital frequency. This can be explained by observing that the quadrupole moment in Eq. (1.92) is invariant under $y^i \to -y^i$. However, this result is only true for the dominant quadrupole radiation. Once higher order correction are included, as shown in Eq. (1.82), harmonics can appear in the form $\omega_n = n\omega$, $n = 1, 2, 3, \ldots$

In reality, the masses do not keep moving on a circular orbit. As orbital energy is radiated away through the emission of GWs, the separation between the two masses shrinks. Consequently, the angular velocity increases causing the system radiate with a larger amplitude and a higher frequency. This behaviour is often referred to as a *chirp* and gives the name to the mass combination in Eq. (1.96).

So far, we have mainly looked at the properties of GWs in the setting of the linearised theory of general relativity GR. However, the full theory of GR is much more complicated compared with this simple, albeit insightful, description. For example, we have ignored how GWs propagates through a curved background, which is described to quadratic order by Eq. (1.62). Furthermore, we have not encountered non-linear aspects of GR such as the scattering of GWs off the background metric. To encounter these non-linear effects, a description of GWs from the *full* EFE is necessary. In Chap. 2, we will discuss a particular framework that is often used to study GWs in the full theory of GR.

Reference

1. C. Misner, K. Thorne, J. Wheeler, *Gravitation* (WH Freeman & co., San Francisco, 1973)

Chapter 2
Gravitational Waves in the Post-Newtonian Formalism

We now turn our attention to GWs in the *full* theory of GR. Solutions to the EFE can only be found analytically in special cases. Instead, approximation methods are commonly used to obtain analytical insight into the full theory of GR.

An example of such an approximation method is the linearised theory described in Chap. 1, where the spacetime metric was assumed to be a flat spacetime metric with some small perturbation and where the source of GWs is modelled through a multipole expansion (or equivalently, an expansion in v/c). However, the linearised theory assumes that the flat background spacetime and the velocity of the sources are somehow independent. For self-gravitating systems, one can no longer assume that the background metric is independent of the velocity of the source. For example, the strength of the gravitational field for a binary system, by equating the gravitational potential to the kinetic energy, is of the order $R_S/R \sim (v/c)^2$, where R_S and R denote the Schwarzschild radius associated to the total mass and the distance between the two masses respectively. Therefore, the aim is to systematically include corrections to linearised theory.

We will give an overview of the post-Newtonian (PN) formalism as developed by Blanchet, Damour, Iyer and co-workers, also known as the BDI formalism. There is a parallel effort by Will, Wiseman and Pati, termed the Direct Integration of the Relaxed Einstein (DIRE) formalism, which can be shown to be completely equivalent to the BDI formalism [1]. For reasons of space, we will only discuss the BDI formalism. Unless explicitly referenced, proofs and more detail can be found in Ref. [2] and references therein.

2.1 Relaxed Einstein Field Equations

To study GWs beyond the linearised theory, it is convenient to first recast the EFE. We define a gravitational field amplitude

$$\mathsf{h}^{\alpha\beta} \equiv \sqrt{-g}\, g^{\alpha\beta} - \eta^{\alpha\beta}, \tag{2.1}$$

T.G.F. Li, *Extracting Physics from Gravitational Waves*,
Springer Theses, DOI 10.1007/978-3-319-19273-4_2

where $g^{\alpha\beta}$, $g \equiv \det(g_{\mu\nu})$ and $\eta^{\alpha\beta}$ denote the metric, its determinant and the Minkowski metric respectively. Similar to Eq. 1.27, we again use the harmonic gauge

$$\mathsf{h}^{\alpha\beta}{}_{,\beta} = 0. \tag{2.2}$$

This newly introduced field amplitude can be, up to a minus sign, shown to reduce to the metric perturbation defined in Eq. (1.25) for small $h_{\alpha\beta}$. Suppose one can write the metric as $g_{\alpha\beta} \approx \eta_{\alpha\beta} + h_{\alpha\beta}$. The gravitational field amplitude in Eq. (2.1) can then be evaluated as

$$\begin{aligned} -\mathsf{h}^{\alpha\beta} &\approx \eta^{\alpha\beta} - \sqrt{1+h}\left(\eta^{\alpha\beta} - h^{\alpha\beta}\right) \\ &\approx h^{\alpha\beta} - \frac{1}{2}\eta^{\alpha\beta} h \\ &= \bar{h}^{\alpha\beta}. \end{aligned} \tag{2.3}$$

In the harmonic gauge, one can rewrite the full EFE in Eq. (1.1) as

$$\Box \mathsf{h}^{\alpha\beta} = 16\pi\tau^{\alpha\beta}, \tag{2.4}$$

where $\Box \equiv \eta^{\mu\nu}\partial_\mu\partial_\nu$ is the flat-space d'Alembertian and

$$\tau^{\alpha\beta} \equiv (-g)T^{\alpha\beta} + \frac{1}{16\pi}\Lambda^{\alpha\beta}, \tag{2.5}$$

$$\begin{aligned} \Lambda^{\alpha\beta} \equiv\; & \mathsf{h}^{\alpha\nu}{}_{,\mu}\mathsf{h}^{\beta\mu}{}_{,\nu} - \mathsf{h}^{\mu\nu}\mathsf{h}^{\alpha\beta}{}_{,\mu\nu} + \frac{1}{2}g^{\alpha\beta}g_{\mu\nu}\mathsf{h}^{\mu\tau}{}_{,\lambda}\mathsf{h}^{\nu\lambda}{}_{,\tau} \\ & - g^{\alpha\mu}g_{\nu\tau}\mathsf{h}^{\beta\tau}{}_{,\lambda}\mathsf{h}^{\nu\lambda}{}_{,\mu} - g^{\beta\mu}g_{\nu\tau}\mathsf{h}^{\alpha\tau}{}_{,\lambda}\mathsf{h}^{\nu\lambda}{}_{,\mu} + g^{\lambda\tau}g_{\mu\nu}\mathsf{h}^{\alpha\mu}{}_{,\lambda}\mathsf{h}^{\beta\nu}{}_{,\tau} \\ & + \frac{1}{8}\left(2g^{\alpha\mu}g^{\beta\nu} - g^{\alpha\beta}g^{\mu\nu}\right)\left(2g_{\lambda\tau}g_{\sigma\rho} - g_{\tau\sigma}g_{\lambda\rho}\right)\mathsf{h}^{\lambda\rho}{}_{,\mu}\mathsf{h}^{\tau\sigma}{}_{,\nu}. \end{aligned} \tag{2.6}$$

The term $\tau^{\alpha\beta}$ can be regarded as the total stress-energy pseudotensor ($\tau^{\alpha\beta}$ is a tensor under Lorentz transformations) that is composed of the matter field, described by $T^{\alpha\beta}$, and the gravitational field, described by $\Lambda^{\alpha\beta}$.

Although Eq. (2.4) bears resemblance to the EFE obtained by using the harmonic gauge in the linearised theory given by Eq. (1.31), it should be stressed that Eq. (2.4) contains no approximations and describes the full theory of GR.

So far, we have rewritten the EFE into two relations Eq. (2.2) and Eq. (2.4). Therefore, just solving for Eq. (2.4) does not yield solutions that also satisfy the EFE. Instead, solutions to the EFE can be obtained by first solving for Eq. (2.4) and then requiring that the condition in Eq. (2.2) holds. Therefore, Eq. (2.4) is referred to as the relaxed EFE. Requiring the harmonic gauge condition implies a conservation law for $\tau^{\alpha\beta}$, given by

$$\tau^{\alpha\beta}{}_{,\beta} = 0, \tag{2.7}$$

which turns out to be equivalent to the conservation law given by Eq. (1.4).

It might seem that one can solve Eq. (2.4), similar to what is done in Sect. (1.6), by imposing the no-incoming-radiation boundary condition and write

$$\mathsf{h}^{\alpha\beta}(t,\vec{x}) = -4\int d^3\vec{y}\,\frac{\tau^{\alpha\beta}\left(t-|\vec{x}-\vec{y}|,\vec{y};\mathsf{h}^{\alpha\beta}\right)}{|\vec{x}-\vec{y}|}. \tag{2.8}$$

However, $\tau^{\alpha\beta}$ is a functional with respect to $\mathsf{h}^{\alpha\beta}$ and a solution to Eq. (2.8) cannot be found trivially. Therefore, approximation methods must be employed in order to gain further insight. Furthermore, since the term $\mathsf{h}^{\alpha\beta}$ is present on the RHS of Eq. (2.8), the resultant GWs will themselves be the source of GWs, a consequence of the non-linear aspect of GR.

2.2 Regions of Interest

Before we continue to explore the relaxed EFE, it is useful to introduce different regions of space in which the gravitational field exhibits different characteristics. This is analogous to the introduction of the near field, far field and the transition region for the electromagnetic field. The PN formalism introduces the *near zone*, which is given by

$$r \ll \lambda\!\!\!^{-}, \tag{2.9}$$

where $\lambda\!\!\!^{-}$ is the reduced wavelength of the GW. In the near zone, retardation effects are negligible. The far zone is given by

$$r \gg \lambda\!\!\!^{-}. \tag{2.10}$$

In contrast to the near zone, retardation effects are important in the far zone. Therefore, different treatments are required for these two regions.

2.2.1 Near Zone

A natural starting point to extend the linearised theory is to look for an expansion of the equations of motion in orders of v/c, a so-called PN expansion. For low velocity sources, one has $\partial_0 = \mathcal{O}(v/c)\partial_i$. Therefore, the d'Alembertian goes as

$$-\partial_0^2 + \nabla^2 = \left[1 + \mathcal{O}\left(\frac{v^2}{c^2}\right)\right]\nabla^2. \tag{2.11}$$

This means that the retardation effects, which are due to the time component of the d'Alembertian, are subdominant and potentials can be viewed as near static. In other words, one tries to expand some quantity $F(t-r)$ for small retardation times, i.e. $t \gg r$. Explicitly, this means that one is looking for an expansion in the form of

$$F(t-r) = F(t) - r\dot{F}(t) + \frac{1}{2}r^2\ddot{F}(t) + \cdots \tag{2.12}$$

Since each derivative carries with it an order of $\omega = 1/\lambda\!\!\!^{-}$, which is the typical GW frequency, the PN expansion can be viewed as an expansion in orders of $r/\lambda\!\!\!^{-}$. Therefore, the PN expansion can only be used in the near zone defined in Eq. (2.9). Indeed, we will explore the use of the PN expansion to describe the near zone in Sect. 2.4.

2.2.2 Far Zone

In the far zone at some distance r, one assumes that the gravitational field is weak and that to some lowest order, spacetime can be described by the Minkowski metric. Corrections to the Minkowski metric will then be given in orders of R_S/r, where again R_S denotes the Schwarzschild radius associated to the total mass of the system. It is convenient to denote the order of the expansion in terms of Newton's constant G, since $R_S \propto G$. Specifically, we write the expansion of the metric as

$$\sqrt{-g}g^{\alpha\beta} = \eta^{\alpha\beta} + G\mathsf{h}_1^{\alpha\beta} + G^2\mathsf{h}_2^{\alpha\beta} + \cdots \tag{2.13}$$

Such an expansion is referred to as the post-Minkowskian (PM) expansion, and is the subject of Sect. 2.3.

2.2.3 Overlap Zone

So far, we have seen that the near zone, for non-relativistic sources ($\lambda\!\!\!^{-} \gg d$), extends up to some radius $\mathfrak{R} \gg d$, where d is the typical size of the system (e.g. orbital radius of a binary). In this near zone, one can use the PN expansion to calculate the gravitational field. The far zone, instead, is taken to be $r \gg \lambda\!\!\!^{-}$. Consider a weak gravitational field inside the object of interest (e.g. a binary system with a large radius). In this case, outside the typical size of the system d, the stress-energy tensor associated to the matter fields vanishes and only the weak gravitational field contributes. As a consequence, the far zone is given by the region $d < r < \infty$. Therefore, there exist, for weakly self-gravitating and non-relativistic sources, a region given by

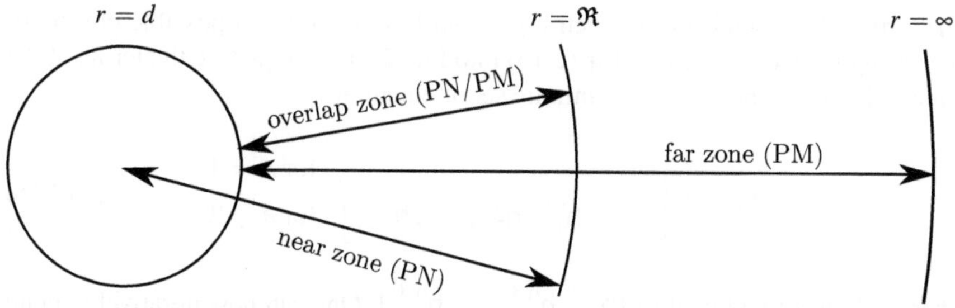

Fig. 2.1 Schematic overview of the different regions and the appropriate expansion considered in the Post-Newtonian formalism for non-relativistic and weakly self-gravitating sources. The near zone corresponds to a region where retardation effects are small, and is given by $0 < r < \mathfrak{R}$, where $\mathfrak{R} \gg d$ is the boundary of the near zone and d the typical size of the source. In the near zone, one can use the Post-Newtonian expansion described in Sect. 2.4. The far zone corresponds to the region where the gravitational fields are weak and is given by $d < r < \infty$. In the far zone, one can use the Post-Newtonian expansion described in Sect. 2.3. For non-relativistic and weakly self-gravitating, an overlap zone exist ($d < r < \infty$) where both PN and PM expansions are valid, as described in Sect. 2.5

$$d < r < \mathfrak{R}, \tag{2.14}$$

where both the PN and the PM expansions are valid. This region is referred to as the *overlap zone* and will prove to be important in connecting the sources of GWs to the GWs observed as $r \to \infty$. A schematic overview of the different regions and their corresponding expansion method is shown in Fig. 2.1.

2.3 Post-Minkowski Expansion in the Far Zone

As discussed in Sect. 2.2.2, the far zone can be described in the PM expansion. In the PM expansion, one expands the field amplitude as a series in powers of $R_S/r \propto G$, as shown in Eq. (2.13). Therefore, we can write

$$\mathsf{h}^{\alpha\beta} = \sum_{n=1}^{\infty} G^n \mathsf{h}_n^{\alpha\beta}. \tag{2.15}$$

Furthermore, we restrict the region of validity to be outside the source so that

$$\tau^{\alpha\beta} = \frac{1}{16\pi}\Lambda^{\alpha\beta}. \tag{2.16}$$

Subsequently, the EFE in Eq. (2.4) becomes

$$\Box \mathsf{h}^{\alpha\beta} = \Lambda^{\alpha\beta}. \tag{2.17}$$

The term $\Lambda^{\alpha\beta}$ depends on the metric $g_{\alpha\beta}$ and has therefore all possible powers of G, starting from G^2. Inserting Eq. (2.15) into Eq. (2.4) and equating the terms of the same order in G, one obtains an infinite set of equations

$$\Box \mathsf{h}_n^{\alpha\beta} = \begin{cases} 0 & \text{for } n = 1 \\ \Lambda_n^{\alpha\beta}\left[\mathsf{h}_1^{\alpha\beta}, \mathsf{h}_2^{\alpha\beta}, \ldots, \mathsf{h}_{n-1}^{\alpha\beta}\right] & \text{for } n \geq 2, \end{cases} \tag{2.18}$$

where $\Lambda_n^{\alpha\beta}$ is a functional of $\{\mathsf{h}_1^{\alpha\beta}, \mathsf{h}_2^{\alpha\beta}, \ldots, \mathsf{h}_{n-1}^{\alpha\beta}\}$. One can now iteratively obtain the solution for $\mathsf{h}_n^{\alpha\beta}$ by solving for h_1 first, and then successively compute higher order terms by using Eq. (2.18) for $n \geq 2$. To retain the correspondence to the full EFE, the solutions must also satisfy the gauge conditions

$$\mathsf{h}_n^{\alpha\beta}{}_{,\beta} = 0. \tag{2.19}$$

2.3.1 Linearised Vacuum Solution

The first step in the PM formalism is to solve the equations for $\mathsf{h}_1^{\alpha\beta}$. These are the so-called linearised vacuum equations

$$\Box \mathsf{h}_1^{\alpha\beta} = 0, \tag{2.20}$$

$$\mathsf{h}_1^{\alpha\beta}{}_{,\beta} = 0. \tag{2.21}$$

The most general solution to Eq. (2.20) for $r > d$, can be expressed in terms of symmetric trace-free (STF) tensors (see Appendix A). For STF tensors, it is convenient to use a multi-index notation due to Blanchet and Damour.

A tensor F with l (spatial) indices $i_1, i_2, \ldots, i_l$ will be compactly written by using $L = i_1 i_2 \ldots i_l$, i.e.

$$F_L \equiv F_{i_1 i_2 \ldots i_l}. \tag{2.22}$$

Similarly, $F_{L-1} = F_{i_1 i_2 \ldots i_{l-1}}$ and $G_{iL} = G_{ii_1 i_2 \ldots i_l}$. Derivative operators can also be compactly written as $\partial_L = \partial_{i_1} \partial_{i_2} \ldots \partial_{i_l}$. Also, we write the product of vectors as

$$x_L \equiv x_{i_1} x_{i_2} \ldots x_{i_l}, \tag{2.23}$$

and reserve $n_i = x_i / r$ as the unit vector in the radial direction. Repeated upper case indices imply the summation over all l indices, i.e.

$$F_L G_L = \sum_{i_1 \ldots i_l} F_{i_1 \ldots i_l} G_{i_1 \ldots i_l}. \tag{2.24}$$

Also, superscripted integers n surrounded by round brackets denote the nth derivative with respect to the retarded time u, i.e.

$$f^{(n)}(u) \equiv d^n f/du^n. \tag{2.25}$$

Round brackets around indices denote the symmetrisation with respect to the enclosed indices, e.g.

$$A_{(ij)} = \frac{1}{2}(A_{ij} + A_{ji}). \tag{2.26}$$

The STF part of a tensor is denoted by either a hat, or by angle brackets surrounding indices, i.e.

$$\hat{A}_L = A_{\langle L \rangle}. \tag{2.27}$$

The use of STF tensors will prove to be useful when one matches the PN expansion to the PM expansion in the overlap zone.

Using the multi-index notation, we most general solution to Eq. (2.20) can be written as

$$\mathsf{h}_1^{\alpha\beta} = \sum_{l=0}^{\infty} \partial_L \left(\frac{K_L^{\alpha\beta}(u)}{r} \right), \tag{2.28}$$

where $r = |\vec{x}|$ and $u \equiv t - r$. If one also imposes Eq. (2.21), the solution can be written as

$$\mathsf{h}_1^{\alpha\beta} = k_1^{\alpha\beta} + \varphi_1^{\beta,\alpha} + \varphi_1^{\alpha,\beta} - \eta^{\alpha\beta}\varphi^{\mu}_{1\,,\mu}, \tag{2.29}$$

where $k_1^{\alpha\beta}$ is given by

$$k_1^{00} = -4 \sum_{l \geq 0} \frac{(-1)^l}{l!} \partial_L \left(\frac{I_L(u)}{r} \right), \tag{2.30}$$

$$k_1^{0i} = 4 \sum_{l \geq 1} \frac{(-1)^l}{l!} \partial_{L-1} \left[\frac{I^{(1)}_{iL-1}(u)}{r} + \frac{l}{l+1} \epsilon_{iab} \partial_a \left(\frac{J_{bL-1}(u)}{r} \right) \right], \tag{2.31}$$

$$k_1^{ij} = -4 \sum_{l \geq 2} \frac{(-1)^l}{l!} \partial_{L-2} \left[\frac{I^{(2)}_{ijL-2}(u)}{r} + \frac{2l}{l+1} \partial_a \left(\frac{\epsilon_{ab(i} J^{(1)}_{j)bL-2}(u)}{r} \right) \right]. \tag{2.32}$$

The tensor $k_1^{\alpha\beta}$ depends on two STF tensors, $I_L(u)$ and $J_L(u)$. From the gauge condition in Eq. (2.21) it follows that I, $I_i^{(1)}$ and J_i are time-independent. This expresses the conservation of total mass, $M \equiv I =$ const, total linear momentum, $P_i \equiv I_i^{(1)} = 0$, and total angular momentum, $S_i \equiv J_i =$ const. The two multipole moments $I_L(u)$ and $J_L(u)$ encode the physical properties of the source at the linearised level, similar to the multipole moments shown in Eq. (1.80), and are referred to as the *mass-type* and *current-type* multipoles.

The remaining terms in Eq. (2.29) depend on φ_1^μ, which is given by

$$\varphi_1^0 = 4\sum_{l\geq 0} \frac{(-1)^l}{l!}\partial_L\left[\frac{W_L(u)}{r}\right], \tag{2.33}$$

$$\begin{aligned}\varphi_1^i = &-4\sum_{l\geq 0} \frac{(-1)^l}{l!}\partial_{iL}\left[\frac{X_L(u)}{r}\right]\\ &-4\sum_{l\geq 1} \frac{(-1)^l}{l!}\partial_{L-1}\left[\frac{Y_{iL-1}(u)}{r} + \frac{l}{l+1}\epsilon_{iab}\partial_a\left(\frac{Z_{bL-1}(u)}{r}\right)\right].\end{aligned} \tag{2.34}$$

These terms originate from the fact that Eq. (2.20) is invariant under linear gauge transformations, $x^\alpha \rightarrow x'^\alpha = x^\alpha + \varphi_1^\alpha$, cf. Eq. (1.11). One might be inclined to simply use the residual gauge freedom to transform away φ_1^μ and recover the results from the linearised theory. However, this cannot be done as $\mathsf{h}_1^{\alpha\beta}$ is used to construct subsequent corrections in $\mathsf{h}^{\alpha\beta}$. Therefore, the set $\{I_L, J_L, W_L, X_L, Y_L, Z_L\}$ is not related to $\{I_L, J_L, 0, 0, 0, 0\}$ through a gauge transformation in the full theory of GR. Instead, one can find a reduced set $\{M_L, S_L, 0, 0, 0, 0\}$ that is related to $\{I_L, J_L, W_L, X_L, Y_L, Z_L\}$ through a gauge transformation. The multipoles M_L and S_L encode the physical properties of the source in the full theory, and differ from I_L and J_L by non-linear corrections.

So far, the set of multipole moments is not characterised in terms of the source $T^{\alpha\beta}$. In fact, this is not possible in the PM regime, as it is intrinsically defined for the region outside the source. However, the source will be characterised in the near zone and the matching procedure will fix the exact expressions for these multipole moments. Before the PN expansion is explored, we first show how to obtain the higher order corrections to $\mathsf{h}_1^{\alpha\beta}$.

2.3.2 Non-linear Iterative Solutions

Once the linearised vacuum solution $\mathsf{h}_1^{\alpha\beta}$ has been found, the next step is to solve

$$\Box \mathsf{h}_n^{\alpha\beta} = \Lambda_n^{\alpha\beta}\left[\mathsf{h}_1^{\alpha\beta}, \mathsf{h}_2^{\alpha\beta}, \ldots, \mathsf{h}_{n-1}^{\alpha\beta}\right], \tag{2.35}$$

$$\mathsf{h}_n^{\alpha\beta}{}_{,\beta} = 0, \tag{2.36}$$

to obtain the higher order corrections. Since the wave equation in Eq. (2.35) is only valid for $r > d$, the retarded integral solution such as Eq. (2.8) is not a solution any more. In fact, the multipole expansion shown in Eqs. (2.30)–(2.34) is singular for $r = 0$.

This problem can be circumvented by realising that one is ultimately interested in the radiation field up to a finite order in the expansion scheme. Consequently, only a finite number of multipoles are assumed to contribute. Therefore, one can find a complex number B, provided the real part of B is large enough, such that $r^B \Lambda_n^{\alpha\beta}$ is well behaved as $r \to 0$. In other words, one is able to construct

$$I_n^{\alpha\beta}(B) = \Box_{\rm ret}^{-1}\left(r^B \Lambda_n^{\alpha\beta}\right), \tag{2.37}$$

where $\Box_{\rm ret}^{-1}$ denotes the convolution with the retarded Green's function, given by

$$(\Box_{\rm ret}^{-1} f)(t, \vec{x}) \equiv -\frac{1}{4\pi}\int d^3\vec{y}\, \frac{f(t - |\vec{x} - \vec{y}|, \vec{y})}{|\vec{x} - \vec{y}|}. \tag{2.38}$$

It can be proven that $I_n^{\alpha\beta}(B)$ admits a unique analytical continuation to all values of B in the complex plane, except for some integers values. When taking $B \to 0$, one can expand $I_n^{\alpha\beta}(B)$ as a Laurent series involving multiple poles

$$I_n^{\alpha\beta} = \sum_{p=p_0}^{\infty} \iota_{np}^{\alpha\beta}(t, \vec{x}) B^p \quad \text{for } p \in \mathbb{Z}, \tag{2.39}$$

where $|p_0|$ is some maximum order of the poles that depends on n. Equating Eq. (2.37) to Eq. (2.39), applying the d'Alembertian to both sides and expanding $r^B = e^{B \log r}$ in terms of B, one can equate the relevant powers of p and obtain

$$\Box \iota_{np}^{\alpha\beta} = \begin{cases} 0 & \text{for } p_0 \le p \le -1, \\ \frac{(\ln r)^p}{p!}\Lambda_n^{\alpha\beta} & \text{for } p \ge 0. \end{cases} \tag{2.40}$$

For the case $p = 0$, one has

$$\Box \iota_{n0}^{\alpha\beta} = \Lambda_n^{\alpha\beta}, \tag{2.41}$$

so that $\iota_{n0}^{\alpha\beta}$, which is called the finite part at $B = 0$ of the retarded integral, turns out to be the particular solution to Eq. (2.35). By defining $u_n^{\alpha\beta} \equiv \iota_{n0}^{\alpha\beta}$ and denoting the finite part at $B = 0$ of the retarded integral as $\mathcal{FP}$, one can write the particular solution as

$$u_n^{\alpha\beta} = \mathcal{FP}\left\{\Box_{\rm ret}^{-1}\left[\Lambda_n^{\alpha\beta}\right]\right\}. \tag{2.42}$$

So far we have only found the particular solution. To obtain the general solution, one must add the solution to the homogeneous equation, $v_n^{\alpha\beta}$. However, the homogeneous solution is simply the solution to the linearised vacuum equation, which is given by Eqs. (2.29)–(2.34). Finally, one can choose $v_n^{\alpha\beta}$ such that $v_n^{\alpha\beta}{}_{,\beta} = -u_n^{\alpha\beta}{}_{,\beta}$ and obtain the general solution to Eqs. (2.35) and (2.36).

2.4 Post-Newtonian Expansion for the Near Zone

So far, we have considered the gravitational fields outside the source by using the PM expansion. However, we have yet to find a way to compute the multipole moments in terms of the stress-energy tensor. Inside the source and within the near field, one can make use of the PN expansion, which is an expansion in powers of v/c.

Similar to the procedure in Sect. 2.3, we start by looking for contributions in successive orders of v/c. Using $1/c^n$ to keep track of the small parameter, the GW field can be written as

$$\mathsf{h}^{\mu\nu} = \sum_{n=2}^{\infty} \frac{1}{c^n} \mathsf{h}_n^{\mu\nu}. \tag{2.43}$$

Similarly, one can write the total stress-energy tensor as

$$\tau^{\mu\nu} = \sum_{n=-2}^{\infty} \frac{1}{c^n} \tau_n^{\mu\nu}. \tag{2.44}$$

Inserting this into the relaxed EFE described in Eq. (2.4) and collecting the relevant powers of v/c, one obtains a set of recursive Poisson-like equations

$$\nabla^2 \mathsf{h}_n^{\mu\nu} = 16\pi \tau_{n-4}^{\mu\nu} + \partial_t^2 h_{n-2}^{\mu\nu}. \tag{2.45}$$

However, these equations cannot be solved by using Green's functions with the boundary condition that there is no incoming radiation. As discussed in Sect. 2.2.1, the PN expansion is essentially an expansion of the retardation effects, which becomes singular as $r \to \infty$.

Nevertheless, one can use a procedure similar to that described in Eqs. (2.37)–(2.42). Instead of having a sufficiently large and positive real part of B, one takes the real part of B to be sufficiently large and negative. Taking the sum over n of both sides of Eq. (2.45) and using the finite part procedure, one finds the particular solution to be

$$u^{\mu\nu} = 16\pi \mathcal{FP} \left\{ \Box_{\text{ret}}^{-1} \left[\tau^{\mu\nu} \right] \right\}. \tag{2.46}$$

Similar to the PM expansion, one assumes that both $u^{\mu\nu}$ and $\tau^{\mu\nu}$ are truncated to a finite order, corresponding to a predetermined accuracy.

Finally, one must add the solution to the homogeneous equation, with the requirement that the solution is regular for $r \to 0$. The homogeneous solution can be shown to have the form

$$v^{\mu\nu} = 16\pi \sum_{l=0}^{\infty} \frac{(-1)^l}{l!} \partial_L \left[\frac{\mathcal{R}_L^{\mu\nu}(t-r) - \mathcal{R}_L^{\mu\nu}(t+r)}{2r} \right], \tag{2.47}$$

where $\mathcal{R}_L^{\mu\nu}(t - r)$ and $\mathcal{R}_L^{\mu\nu}(t + r)$ are arbitrary functions of $t - r$ and $t + r$ respectively. This follows again from $\Box\left[f(t-r)/r\right] = 0$ and $\Box\left[g(t+r)/r\right] = 0$ for arbitrary functions $f(t-r)$ and $g(t+r)$. The functions $\mathcal{R}_L^{\mu\nu}$ will be used later when one matches the PN expansion to the PM expansions in the overlap zone.

2.5 Matching of the Post-Minkowski and Post-Newtonian Expansions

We can now take stock of what has been done so far. In Sect. 2.3 we introduced the PM expansion which is valid in the domain $d < r < \infty$. This PM expansion is supplemented with the multipolar expansion from Sect. 2.3 and Appendix A. The most general solution in the PM expansion is characterised by the set of multipole moments $\{I_L, J_L, W_L, X_L, Y_L, Z_L\}$. However, the problem is that this set of multipole moments is unspecified.

Next, we introduced the PN expansion in Sect. 2.4. This expansion is only valid in the near-zone regime $0 < r < \mathcal{R}$, where $\mathcal{R} \gg \lambda\!\!\!^{-}$. In the PN expansion, we obtain the GW field as a function of the stress-energy tensor of the source.

Provided that source is non-relativistic and weakly self-gravitating, one can deduce that an overlap zone $d < r < \mathcal{R}$ exist where both the multipolar PM and the PN expansion hold. Suppose we denote the multipolar expansion of h with $\mathcal{M}(\mathsf{h})$ and the PN expansion with $\mathcal{N}(\mathsf{h})$. Then, within the overlap zone, one can write down the so-called *matching condition*

$$\mathcal{N}(\mathcal{M}(\mathsf{h})) = \mathcal{M}(\mathcal{N}(\mathsf{h})). \tag{2.48}$$

In other words, one can re-expand the multipolar PM expansion as a PN expansion and relate that to the re-expansion of the PN expansion in terms of a multipolar expansion. With the matching condition, one can express the set of multipole moments $\{I_L, J_L, W_L, X_L, Y_L, Z_L\}$ in Eqs. (2.30)–(2.34) for the PM expansion in terms of the total stress-energy tensor $\tau^{\mu\nu}$ in Eq. (2.46) for the PN expansion. Specifically, the multipole moments can written as

$$I_L(u) = \mathcal{FP}\int d^3\vec{y}\int_{-1}^{1} dz\left\{\delta_l \hat{y}_L \Sigma - \frac{4(2l+1)\delta_{l+1}}{(l+1)(2l+3)}\hat{y}_{iL}\Sigma_i^{(1)} + \frac{2(2l+1)\delta_{l+2}}{(l+1)(l+2)(2l+5)}\hat{y}_{ijL}\Sigma_{ij}^{(2)}\right\}(u+z|\vec{y}|,\vec{y}), \tag{2.49}$$

$$J_L(u) = \mathcal{FP}\int d^3\vec{y}\int_{-1}^{1} dz\; \epsilon_{ab\langle i_l}\left\{\delta_l \hat{y}_{L-1\rangle a}\Sigma_b - \frac{(2l+1)\delta_{l+1}}{(l+2)(2l+3)}\hat{y}_{L-1\rangle ac}\Sigma_{bc}^{(1)}\right\}(u+z|\vec{y}|,\vec{y}), \tag{2.50}$$

$$W_L(u) = \mathcal{FP}\int d^3\vec{y}\int_{-1}^{1} dz\left\{\frac{(2l+1)\delta_{l+1}}{(l+1)(2l+3)}\hat{y}_{iL}\Sigma_i - \frac{(2l+1)\delta_{l+2}}{2(l+1)(l+2)(2l+5)}\hat{y}_{ijL}\Sigma_{ij}^{(1)}\right\}(u+z|\vec{y}|,\vec{y}), \tag{2.51}$$

$$X_L(u) = \mathcal{FP}\int d^3\vec{y}\int_{-1}^{1} dz\left\{\frac{(2l+1)\delta_{l+2}}{2(l+1)(l+2)(2l+5)}\hat{y}_{ijL}\Sigma_{ij}\right\}(u+z|\vec{y}|,\vec{y}), \tag{2.52}$$

$$Y_L(u) = \mathcal{FP}\int d^3\vec{y}\int_{-1}^{1} dz\left\{-\delta_l \hat{y}_L \Sigma_{ii} + \frac{3(2l+1)\delta_{l+1}}{(l+1)(2l+3)}\hat{y}_{iL}\Sigma_i^{(1)} - \frac{2(2l+1)\delta_{l+2}}{(l+1)(l+2)(2l+5)}\hat{y}_{ijL}\Sigma_{ij}^{(2)}\right\}(u+z|\vec{y}|,\vec{y}), \tag{2.53}$$

$$Z_L(u) = \mathcal{FP}\int d^3\vec{y}\int_{-1}^{1} dz\; \epsilon_{ab\langle i_l}\left\{-\frac{(2l+1)\delta_{l+1}}{(l+2)(2l+3)}\hat{y}_{L-1\rangle bc}\Sigma_{ac}\right\}(u+z|\vec{y}|,\vec{y}), \tag{2.54}$$

where Σ, Σ_i and Σ_{ij} are given by

$$\Sigma = \tau^{00} + \tau^{ii}, \quad \Sigma_i = \tau^{0i}, \quad \Sigma_{ij} = \tau^{ij}. \tag{2.55}$$

These multipole moments are to be complemented with the functions $\mathcal{R}_L^{\mu\nu}$ appearing in the homogeneous solution in Eq. (2.47). These functions are chosen such that the matching condition in Eq. (2.48) is satisfied. Specifically, the result of this choice is given by

$$\mathcal{R}_L^{\mu\nu}(u) = \frac{1}{2\pi}\mathcal{FP}\int d^3\vec{y}\;\hat{y}_L\int_1^{\infty} dz\;\delta_l \mathcal{M}\left(\tau^{\mu\nu}\right)(u - z|\vec{y}|,\vec{y}). \tag{2.56}$$

2.6 Waveforms for Binary Systems

With the PN formalism described in Sects. 2.1–2.5, one can model GWs emitted by various sources. In this work, we will mainly consider GWs emitted by binary systems (see Fig. 1.2), for which it will be useful to define the dimensionless frequency variable

$$x \equiv \left(\frac{GM\omega}{c^3}\right)^{2/3}, \tag{2.57}$$

where $M = m_1 + m_2$ is the total mass and $\omega = 2\pi f$ is the angular velocity. This variable can be shown to be $x \sim \mathcal{O}(1/c^2)$ and is useful to keep track of the relevant orders in the expansion. Furthermore, we adopt the nomenclature of the nPN order for terms proportional to x^n. For example, the 3.5PN order refers to the terms proportional to $x^{7/2} \sim \mathcal{O}(1/c^7)$. For clarity, we will reinstate the factors of c and G. Furthermore, we define the so-called *symmetric mass ratio*

$$\eta \equiv \frac{\mu}{M} = \frac{m_1 m_2}{(m_1 + m_2)^2}, \tag{2.58}$$

where $\mu = m_1 m_2 / M$ is the reduced mass. The post-Newtonian parameter, which is useful to keep track of successive PN corrections, is defined as

$$\gamma \equiv \frac{GM}{Rc^2} = \mathcal{O}\left(\frac{1}{c^2}\right), \tag{2.59}$$

where R is the separation between the two masses. Finally, we introduce a dimensionless time variable

$$\Theta \equiv \frac{\eta c^3}{5GM}(t_c - t), \tag{2.60}$$

where t_c represents the time at which the two objects coalesce.

2.6.1 Equations of Motion

The equations of motion are obtained from the geodesic equation given by Eq. (1.5) together with the metric obtained from the PN formalism. We denote the relative separation between the two masses in the centre of mass frame by $y^i = y_1^i - y_2^i$, where y_1^i and y_2^i are the positions of masses 1 and 2. The relative velocity is given by $v^i = v_1^i - v_2^i$, where $v_1^i = dy_1^i/dt$ is the velocity of mass 1 (and similar for mass 2). The equations of motion are given by

$$\frac{dv^i}{dt} = -\frac{GM}{R^2}\left\{\left[1 + \mathcal{A}(R, \eta, M, v^i)\right] n^i + \mathcal{B}(R, \eta, M, v^i) v^i\right\} + \mathcal{O}\left(\frac{1}{c^8}\right), \tag{2.61}$$

where $n^i = y^i/R$ denotes the unit direction between the two masses. The coefficients $\mathcal{A}$ and $\mathcal{B}$ are complicated expressions of R, η, M and v^i. This expression can be simplified by noting that the radiation reaction tends to circularise the orbit and that by the time such signals are detectable by Earth-based detectors, a circular orbit is an adequate representation. For circular orbits, Eq. (2.61) simplifies to

$$\frac{dv^i}{dt} = -\omega^2 y^i - \frac{32}{5}\frac{G^3M^3\eta}{c^5R^4}v^i + \mathcal{O}\left(\frac{1}{c^7}\right). \tag{2.62}$$

The orbital frequency is given by the generalised version of Kepler's third law

$$\omega^2 = \frac{GM}{r^3}\left\{1 + (-3+\eta)\gamma + \left(6 + \frac{41}{4}\eta + \eta^2\right)\gamma^2 + \left(-10 + \left[-\frac{75707}{840} - \frac{41}{64}\pi^2 + 22\ln\left(\frac{R}{R_0'}\right)\right]\eta + \frac{19}{2}\eta^2 + \eta^3\right)\gamma^3\right\} + \mathcal{O}\left(\frac{1}{c^8}\right), \tag{2.63}$$

where the length scale R_0' is a gauge-related constant. In fact, R is the separation between the two masses in the harmonic-gauge coordinates, and is therefore not invariant under a coordinate transformation. However GR requires that all physical observables are invariant under a coordinate transformation, and we will see that within the PN formalism this is still the case.

From Eq. (2.61), one can deduce that the orbital energy of the circular orbit is given by

$$E = -\frac{\mu c^2\gamma}{2}\left\{1 + \left(-\frac{7}{4} + \frac{1}{4}\eta\right)\gamma + \left(-\frac{7}{8} + \frac{49}{8}\eta + \frac{1}{8}\eta^2\right)\gamma^2 + \left[-\frac{235}{64} + \left(\frac{46031}{2240} - \frac{123}{64}\pi^2 + \frac{22}{3}\ln\left(\frac{R}{R_0'}\right)\right)\eta + \frac{27}{32}\eta^2 + \frac{5}{64}\eta^3\right]\gamma^3\right\} + \mathcal{O}\left(\frac{1}{c^8}\right). \tag{2.64}$$

This expression still seems to have terms that depend on the specific coordinate system through γ and R_0'. However, one can rewrite this expression in terms of x from Eq. (2.57) by solving Eq. (2.63) for γ as a function of x. The resulting expression is given by

$$E = -\frac{\mu c^2 x}{2}\left\{1 + \left(-\frac{3}{4} - \frac{1}{12}\eta\right)x + \left(-\frac{27}{8} + \frac{19}{8}\eta - \frac{1}{24}\eta^2\right)x^2 + \left[-\frac{675}{64} + \left(\frac{34445}{576} - \frac{205}{96}\pi^2\right)\eta - \frac{155}{96}\eta^2 - \frac{35}{5184}\eta^3\right]x^3\right\} + \mathcal{O}\left(\frac{1}{c^8}\right), \tag{2.65}$$

from which we see that all the coordinate dependence has indeed cancelled out. Finally, the energy of a circular orbit is calculated to 3.5PN order. This is the highest order to which the energy of a binary system is known. The expressions for the full non-circular binary system can be found in Ref. [2].

2.6.2 Energy Flux

The computation of the equations of motion in the near zone is only one part of the output of the PN formalism. The other part is the computation of the waveform as $r \to \infty$. From this waveform, one can determine the energy flux as $r \to \infty$. In particular, the total flux from GWs is obtained from the binary's multipole moments, similar to Eq. (1.88), and is given by

$$\begin{aligned}\mathcal{L} = \frac{32c^5}{5G}\eta^2 x^5 \Bigg\{ 1 &+ \left(-\frac{1247}{336} - \frac{35}{12}\eta\right) x + 4\pi x^{3/2} + \left(-\frac{44711}{9072} + \frac{9271}{504}\eta + \frac{65}{18}\eta^2\right) x^2 \\ &+ \left(-\frac{8191}{672} - \frac{583}{24}\eta\right)\pi x^{5/2} + \left[\frac{6643739519}{69854400} + \frac{16}{3}\pi^2 - \frac{1712}{105}C - \frac{856}{105}\ln(16x)\right. \\ &+ \left.\left(-\frac{134543}{7776} + \frac{41}{48}\pi^2\right)\eta - \frac{94403}{3024}\eta^2 - \frac{775}{324}\eta^3\right] x^3 \\ &+ \left(-\frac{16285}{504} + \frac{214745}{1728}\eta + \frac{193385}{3024}\eta^2\right)\pi x^{7/2} + \mathcal{O}\left(\frac{1}{c^8}\right)\Bigg\}, \end{aligned} \tag{2.66}$$

where $C = 0.577\ldots$ is the Euler-Mascheroni constant.

2.6.3 Waveform

Orbital Phase

One can now relate the orbital energy given by Eq. (2.65) to the flux described by Eq. (2.66) through the relationship

$$\frac{dE}{dt} = -\mathcal{L}. \tag{2.67}$$

Although this relationship seems to be physically intuitive, no proof from first principles of GR exist for this relationship at 3PN accuracy. However, this equation has been checked up to 1.5PN, which does include non-linear contributions. Nevertheless, we will assume the validity of Eq. (2.67) to arbitrary PN orders and use it to obtain the temporal evolution of the orbital frequency, i.e.

$$\begin{aligned} x = \frac{1}{4}\Theta^{-1/4} \Bigg\{ 1 &+ \left(\frac{743}{4032} + \frac{11}{48}\eta\right)\Theta^{-1/4} - \frac{1}{5}\pi\Theta^{-3/8} + \left(\frac{19583}{254016} + \frac{24401}{193536}\eta\right. \\ &+ \left.\frac{31}{288}\eta^2\right)\Theta^{-1/2} + \left(-\frac{11891}{53760} + \frac{109}{1920}\eta\right)\pi\Theta^{-5/8} + \left[-\frac{10052469856691}{6008596070400} + \frac{1}{6}\pi^2 + \frac{107}{420}C\right. \\ &- \left.\frac{107}{3360}\ln\left(\frac{\Theta}{256}\right) + \left(\frac{3147553127}{780337152} - \frac{451}{3072}\pi^2\right)\eta - \frac{15211}{442368}\eta^2 + \frac{25565}{331776}\eta^3\right]\Theta^{-3/4} \\ &+ \left(-\frac{113868647}{433520640} - \frac{31821}{143360}\eta + \frac{294941}{3870720}\eta^2\right)\pi\Theta^{-7/8} + \mathcal{O}\left(\frac{1}{c^8}\right)\Bigg\}. \end{aligned} \tag{2.68}$$

The accumulated orbital phase can be found by integrating the orbital frequency

$$\phi(t) = \int_{t_0}^{t} dt' \omega(t'). \tag{2.69}$$

After performing the integration, the orbital phase as a function of Θ is given by

$$\begin{aligned}\phi = &-\frac{1}{\eta}\Theta^{5/8}\Bigg\{1+\left(\frac{3715}{8064}+\frac{55}{96}\eta\right)\Theta^{-1/4}-\frac{3}{4}\pi\Theta^{-3/8}+\left(\frac{9275495}{14450688}+\frac{284875}{258048}\eta\right.\\ &\left.+\frac{1855}{2048}\eta^2\right)\Theta^{-1/2}+\left(-\frac{38645}{172032}+\frac{65}{2048}\eta\right)\pi\Theta^{-5/8}\ln\left(\frac{\Theta}{\Theta_0}\right)\\ &+\left[\frac{831032450749357}{57682522275840}-\frac{53}{40}\pi^2-\frac{107}{56}C+\frac{107}{448}\ln\left(\frac{\Theta}{256}\right)\right.\\ &\left.+\left(-\frac{126510089885}{4161798144}+\frac{2255}{2048}\pi^2\right)\eta+\frac{154565}{1835008}\eta^2-\frac{1179625}{1769472}\eta^3\right]\Theta^{-3/4}\\ &+\left(\frac{188516689}{173408256}+\frac{488825}{516096}\eta-\frac{141769}{516096}\eta^2\right)\pi\Theta^{-7/8}+\mathcal{O}\left(\frac{1}{c^8}\right)\Bigg\},\end{aligned} \tag{2.70}$$

where Θ_0 is a constant of integration that can be fixed by the initial conditions, e.g. when the GW frequency enters the detector's bandwidth. Finally, the phase can also be written as a function of x

$$\begin{aligned}\phi = &-\frac{x^{-5/2}}{32\eta}\Bigg\{1+\left(\frac{3715}{1008}+\frac{55}{12}\eta\right)x-10\pi x^{3/2}+\left(\frac{15293365}{1016064}+\frac{27145}{1008}\eta+\frac{3085}{144}\eta^2\right)x^2\\ &+\left(\frac{38645}{1344}-\frac{65}{16}\eta\right)\pi x^{5/2}\ln\left(\frac{x}{x_0}\right)+\left[\frac{12348611926451}{18776862720}-\frac{160}{3}\pi^2-\frac{1712}{21}C\right.\\ &\left.-\frac{856}{21}\ln(16x)+\left(-\frac{15737765635}{12192768}+\frac{2255}{48}\pi^2\right)\eta+\frac{76055}{6912}\eta^2-\frac{127825}{5184}\eta^3\right]x^3\\ &+\left(\frac{77096675}{2032128}+\frac{378515}{12096}\eta-\frac{74045}{6048}\eta^2\right)\pi x^{7/2}+\mathcal{O}\left(\frac{1}{c^8}\right)\Bigg\},\end{aligned} \tag{2.71}$$

where x_0 is an integration constant similar to Θ_0.

Full Waveform

The full waveform is currently known up to 3PN order [3]. With the same labelling and coordinate system as Sect. 1.6.2, the two polarisations can be written as

$$h_{+,\times} = \frac{2G\mu x}{c^2 r}\left\{H^{(0)}_{+,\times}+x^{1/2}H^{(1/2)}_{+,\times}+xH^{(1)}_{+,\times}+\mathcal{O}\left(\frac{1}{c^3}\right)\right\}. \tag{2.72}$$

The leading terms of $H_{+,\times}$ are given by

$$H^{(0)}_{+} = -(1+c_i^2)\cos 2\psi, \tag{2.73}$$

$$H^{(0)}_{\times} = -2c_i \sin 2\psi, \tag{2.74}$$

$$H_+^{(1/2)} = -\frac{s_i}{8}\frac{\delta m}{m}\left[(5+c_i^2)\cos\psi - 9(1+c_i^2)\cos 3\psi\right], \tag{2.75}$$

$$H_\times^{(1/2)} = -\frac{3}{4}s_i c_i \frac{\delta m}{m}\left[\sin\psi - 3\sin 3\psi\right], \tag{2.76}$$

$$H_+^{(1)} = \frac{1}{6}\left[19 + 9c_i^2 - 2c_i^4 - \eta(19 - 11c_i^2 - 6c_i^4)\right]\cos 2\psi$$
$$- \frac{4}{3}s_i^2(1+c_i^2)(1-3\eta)\cos 4\psi, \tag{2.77}$$

$$H_\times^{(1)} = \frac{c_i}{3}\left[17 - 4c_i^2 - \eta(13 - 12c_i^2)\right]\sin 2\psi - \frac{8}{3}(1-3\eta)c_i s_i^2 \sin 4\psi, \tag{2.78}$$

where $c_i = \cos\iota$, $s_i = \sin\iota$ and $\delta m = m_1 - m_2$. The auxiliary phase is defined as

$$\psi = \phi - \frac{2GM\omega}{c^3}\ln\left(\frac{\omega}{\omega_0}\right), \tag{2.79}$$

where ω_0 is the frequency at which the signal enters the sensitive region of the detector. Remarkably, to leading order in amplitude, one recovers the form obtained from the linear theory given by Eq. (1.93) and Eq. (1.94),[1] only differing by the definition of the phase. Furthermore, the inclusion of higher order multipoles indeed leads to harmonics at $n\psi$, where $n = 1, 2, 3, \ldots$.

2.6.4 Additional Contributions

In the calculations of the waveform, we have assumed that the components making up the binary are non-spinning and point-like. Furthermore, we have assumed that the orbit of the binary is quasi-circular. In this section, we will look into some of the modifications required to relax these assumptions.

Spin

Astrophysical observations have suggested that black holes can be spinning significantly [4–6]. To introducing spin into the model, one requires the inclusion of the spin-orbit and spin-spin coupling. In other words, to model these additional interactions requires a modification to the equation of motion shown in Eq. (2.61) and therefore the orbital energy shown in Eq. (2.65) (see Ref. [7] and references therein).

The result is that the waveform of a binary is augmented with spin-orbit terms starting from the 1.5PN order, and with spin-spin terms starting from the 2PN order. The spin couplings introduce rich dynamics resulting in the modulation of the amplitude and the frequency of the waveform. However, the inclusion of spin couplings requires an additional six parameters to describe the waveform. This in turn will add complexity to data-analysis tasks such as testing the strong field dynamics of GR.

[1] Assuming Kepler's third law for circular orbits, i.e. $R^3 = GM/\omega^2$.

Extended Bodies

So far, we have been treating the objects in binaries as point-like objects with a certain mass. Such an approximation is valid when the objects are far away from each other. However, as the two objects more closer together due to the loss of energy caused by the emission of GWs, the internal structure of the bodies start to have an influence on the waveform.

Tidal Deformability Consider a star with mass m that experiences the quadrupolar tidal field $\mathcal{E}_{ij}$ of its companion. This quadrupolar tidal field will then induce a quadrupole moment Q_{ij} in the star [8]. To lowest order in $\mathcal{E}_{ij}$, one can relate the induced quadrupolar field to the external tidal field through

$$Q_{ij} = -\lambda(m)\mathcal{E}_{ij}. \tag{2.80}$$

The function $\lambda(m)$ is called the tidal deformability and depends on the internal characteristics of the star (for black holes, $\lambda = 0$). It turns out that the corrections related to the star's tidal deformability start from the 5PN order. However, the prefactor associated with these terms is of such magnitude that the measurement of these terms are in fact plausible and can provide us with information about the internal structure of the object [9].

Oblateness Caused by Rotating Stars In addition to the distortion caused by an external tidal field, extended objects can also acquire further distortion through the rotation of the star itself. This process will introduce a quadrupole term in the gravitational potential and modify the equations of motion. The waveform modification starts from the 2PN order and depends on the spin, and the internal structure through the quadrupole moment scalar $Q(m)$ (not to be confused with the quadrupole moment in Eq. (2.80)) [10]. Similar to the tidal deformability, this spin-induced quadrupolar moment also depends on the internal structure of the object.

I-Love-Q Relationship It was recently suggested that there might be a phenomenological relationship between the tidal deformability λ and the spin-induced quadrupole moment scalar Q [11–13]. Such a relationship will be useful to connect the tidal deformability to the spin-induced quadrupole moment scalar and provide additional information about the internal structure of the object.

Eccentricity

Most binary systems are born in bound systems. As a consequence, the system has enough time to circularise the orbit due to the emission of GWs [14, 15], before the GW signal becomes detectable by Earth-based detectors. Therefore, most binaries are considered to be in quasi-circular orbits when their signal reaches the sensitive region of the detector. However, in dense stellar regions such as the cores of galaxies or globular clusters, stars can become gravitationally bound to a companion if their trajectories are close enough [16–18]. In this case, the orbit cannot be assumed to be quasi-circular, and one has to model GWs from binaries in eccentric orbits.

The effect of eccentric orbits on the waveform is an interplay of many effects. For example, eccentric binaries will cause additional bursts of GW emission near the periastron passage and also a shift in the frequency of the dominant harmonic (frequency increases as the eccentricity increases) [19]. However, the bursts of GW emission near periastron passage will cause the circularisation of the orbit and therefore weakening the effects of eccentric orbits over time.

Merger and Ringdown

So far, we have considered the PN formalism from which follows that the waveform for binary systems can be written as an expansion in v/c. However, as the binary loses energy and angular momentum through the emission of GWs, the orbital radius shrinks and the frequency (hence the typical velocity) of the binary goes up. Therefore, there will be a point where the PN formalism breaks down.

The break-down of the PN roughly coincides with the situation that the two objects are sufficiently close together so that they start to merge into a single object. During this stage of the binary evolution, the system is highly relativistic and self-gravitating. No method has yet to be found to model this stage analytically, and one has to resort to solving the EFE numerically instead.

Once the two objects have merged and formed a single object, it will further emit GWs as the object tries to reach a quiescent state. The GWs emitted during this phase can be described by a superposition of damped sinusoids, also referred to as *quasi-normal modes*. As a consequence of the no-hair theorem, the oscillation frequencies and the associated damping times of these quasi-normal modes are a function of the mass and the spin only.

2.6.5 Validity of the Post-Newtonian Formalism

Since the PN formalism is merely an approximation method, it is natural to wonder how accurate such a scheme is in describing the GWs emitted by binary systems. This question is difficult to answer for several reasons. Firstly, by the nature of any approximation method, it is difficult to quantify the higher orders that have yet to be calculated. Secondly, the development of approximation methods is necessary as numerical methods are computationally expensive and challenging. Modern computers still cannot simulate the evolution of the binary over a few tens, or produce the high number of waveforms necessary to systematically asses the validity of approximation methods such as the PN formalism.

A clue to the accuracy of the PN formalism is the determination of the innermost circular orbit, which is defined as the minimum of the energy given by Eq. (2.64) or Eq. (2.65). It was shown that the PN formalism up to the 3PN order estimates the frequency and the energy at the innermost circular orbit to within a fractional accuracy of about 1 % [20].

Furthermore, comparisons between waveforms from numerical simulations and the PN formalism show that close to the merger of the two objects, assuming an equal starting point, the difference is only a few percent over a time-span of six cycles [21]. The disagreements between these two methods are thought to be caused by the truncation of the expansion series in the PN formalism.

It is common in the literature to assume the validity of the PN formalism up to the point of the last stable orbit (LSO) indexlast stable orbit, which in the test particle limit occurs at separation of $R = 6M$ and has $v/c = 1/\sqrt{6}$. In some cases, the frequency evolution stops being monotonic before LSO indexlast stable orbit, which is another indicator that the PN formalism breaks down and the waveform should no longer be trusted. The actual truncation of the waveform depends the specific implementation.

2.6.6 Stationary Phase Approximation

For data-analysis purposes, as we will see in Sect. 3.2, it is important to use the Fourier transform of the time-domain waveform shown in Eq. (2.72). However, the Fourier transform is difficult to compute analytically. Instead, one could perform the Fourier transform numerically. Since the waveform has to be calculated many times over for different values of the associated parameters, performing the Fourier transform numerically proves to be computationally expensive. Luckily, one can use the SPA to get an approximation of the Fourier transform of a time-domain waveform.

Consider a waveform at the 0PN order in amplitude. The plus and cross polarisation of the GW can be written in a generic time domain form given by

$$h_{+,\times}(t) = A_{+,\times}(t) \cos \Phi_{+,\times}(t). \tag{2.81}$$

The goal is then to compute the Fourier transform of Eq. (2.81), given by

$$\begin{aligned} \tilde{h}_{+,\times}(f) &= \int_{-\infty}^{\infty} dt \; A_{+,\times}(t) \cos \Phi_{+,\times}(t) e^{2\pi \imath f t} \\ &= \frac{1}{2} \int_{-\infty}^{\infty} dt \; A_{+,\times}(t) \left(e^{\imath (2\pi f t + \Phi_{+,\times}(t))} + e^{\imath (2\pi f t - \Phi_{+,\times}(t))} \right). \end{aligned} \tag{2.82}$$

The idea of the SPA is to evaluate the integral in Eq. (2.82) only around the point where one expects the biggest contribution. Explicitly, this means that one neglects the first term of Eq. (2.82) because it always oscillates rapidly and averages to zero. The second term also approximately averages to zero, except around the point t_*, where

$$2\pi f = \dot{\Phi}_{+,\times}(t_*) = \pi f_{\mathrm{GW}}(t_*). \tag{2.83}$$

Next, we can expand the phase up to $(t - t_*)^2$, and make an additional approximation that the amplitude also varies slowly around t_*, allowing us to take the amplitude out of the integral. The GW field can then be written as

$$\tilde{h}_{+,\times} = \frac{1}{2} A_{+,\times}(t_*) e^{\imath(2\pi f t_* - \Phi_{+,\times}(t_*))} \sqrt{\frac{2}{\ddot{\Phi}_{+,\times}(t_*)}} \int_{-\infty}^{\infty} dx\, e^{-\imath x^2}, \tag{2.84}$$

where $x = \sqrt{\Phi_{+,\times}(t_*)/2}$. The remaining integral can be evaluated [22] as

$$\int_{-\infty}^{\infty} dx\, e^{-\imath x^2} = \sqrt{\pi} e^{-\imath\pi/4}. \tag{2.85}$$

Therefore, Eq. (2.84) becomes

$$\tilde{h}_{+,\times} = \frac{1}{2} A_{+,\times}(t_*) \sqrt{\frac{2\pi}{\ddot{\Phi}_{+,\times}(t_*)}} e^{-\imath\Psi_{+,\times}}, \tag{2.86}$$

where the phase $\Psi_{+,\times}$ is given by

$$\Psi_{+,\times} = 2\pi f t_* - \Phi_{+,\times}(t_*) - \frac{\pi}{4}. \tag{2.87}$$

Finally, one can invert the expression in Eq. (2.68) to eliminate t_* in favour of f.

The phase in the SPA has been calculated up to the 3.5PN order [23–25], and is given by

$$\Psi_{+,\times}(f) = 2\pi f t_c - \phi_c - \frac{\pi}{4} + \sum_{j=0}^{7} \left[\psi_j + \psi_j^{(l)} \ln f\right] f^{(j-5)/3}, \tag{2.88}$$

where t_c and ϕ_c are the time and phase of coalescence. The phase coefficients ψ_j and $\psi_j^{(l)}$ are given by

$$\psi_j = \frac{3}{128\,\eta} (\pi M)^{(j-5)/3} \alpha_j \quad \text{and} \quad \psi_j^{(l)} = \frac{3}{128\,\eta} (\pi M)^{(j-5)/3} \alpha_j^{(l)}, \tag{2.89}$$

and the α_j coefficients are given by

$$\alpha_0 = 1, \tag{2.90}$$

$$\alpha_1 = 0, \tag{2.91}$$

$$\alpha_2 = \frac{3715}{756} + \frac{55}{9}\eta, \tag{2.92}$$

$$\alpha_3 = -16\pi, \tag{2.93}$$

$$\alpha_4 = \frac{15293365}{508032} + \frac{27145}{504}\eta + \frac{3085}{72}\eta^2, \tag{2.94}$$

$$\alpha_5 = \pi\left(\frac{38645}{756} - \frac{65}{9}\eta\right)\left[1 + \ln\left(2\,6^{3/2}\pi M\right)\right], \tag{2.95}$$

$$\alpha_6 = \frac{11583231236531}{4694215680} - \frac{640}{3}\pi^2 - \frac{6848}{21}C + \left(-\frac{15737765635}{3048192} + \frac{2255}{12}\pi^2\right)\eta$$
$$+ \frac{76055}{1728}\eta^2 - \frac{127825}{1296}\eta^3 - \frac{6848}{63}\ln(128\,\pi M), \tag{2.96}$$
$$\alpha_7 = \pi\left(\frac{77096675}{254016} + \frac{378515}{1512}\eta - \frac{74045}{756}\eta^2\right), \tag{2.97}$$
$$\alpha_j^{(l)} = 0, \quad \text{for } j = 0, 1, 2, 3, 4, 7, \tag{2.98}$$
$$\alpha_5^{(l)} = \pi\left(\frac{38645}{756} - \frac{65}{9}\eta\right), \tag{2.99}$$
$$\alpha_6^{(l)} = -\frac{6848}{63}. \tag{2.100}$$

The amplitude to the 0PN order is determined by $A(t_*)$ and $\ddot{\Phi}(t_*)$. The expression from $A(t_*)$ is obtained from Eq. (2.73) and the inversion of Eq. (2.68) to eliminate t_* in favour of f. Following Eq. (2.83), one can write $\ddot{\Phi}(t_*) = \pi \dot{f}_{\rm GW}(t_*)$, where $\dot{f}_{\rm GW}$ can be evaluated as

$$\begin{aligned}\dot{f}_{\rm GW}(t) &= \frac{df_{\rm GW}}{dt} \\ &= \frac{df_{\rm GW}}{dx}\frac{dx}{dE}\frac{dE}{dt} \\ &= \frac{3}{2}\frac{x}{\pi M}\frac{-\mathcal{L}(x)}{E'(x)}.\end{aligned} \tag{2.101}$$

The prime denotes a derivative with respect to x. The SPA of the waveform at the 0PN order in amplitude is referred to as the *TaylorF2* waveform in the literature (e.g. [26]).

Accuracy of the Stationary Phase Approximation

A recent study has compared the SPA to time-domain waveforms from the PN formalism [26]. It was shown that the SPA agrees with time-domain waveforms with an accuracy between a fraction of a percent to a few percent depending on the total mass of the system (the SPA becomes less accurate compared to the time-domain waveform as the total mass goes up).

2.6.7 Concluding Remarks

The waveform emitted by binary systems depends intricately on the intrinsic parameters such as the masses and the spin, as well as the extrinsic parameters such as the distance and the orbital orientation. Measuring GWs emitted by binary system (see Chap. 3) means that one can infer these parameters (see Chaps. 4 and 5) and possibly obtain some answers to astrophysical problems.

Two of such problems will be the focus of this book. Firstly, the functional dependence of the phase coefficients in Eq. (2.100) on the component masses is set by GR. As will be shown in part II, these phase coefficients can be used to construct a generic test of GR. Secondly, as shown in Eq. (2.72), the amplitude of the waveform is proportional to $1/r$, where r is the distance between the observer and the source. This distance can be related to the luminosity distance. Therefore, one can use GWs emitted by binary systems to measure the luminosity distance directly, and subsequently obtain cosmological information. This will be the topic of Part III.

References

1. M. E. Pati, C. M. Will, Post-Newtonian gravitational radiation and equations of motion via direct integration of the relaxed Einstein equations: foundations. Phys. Rev. D **62**(12), 124015 (2000)
2. L. Blanchet, Gravitational radiation from post-newtonian sources and inspiralling-compact binaries. Living Rev. Relativ. **9**(4) (2006)
3. L. Blanchet et al., The third post-newtonian gravitational wave polarizations and associated-spherical harmonic modes for inspiralling compact binaries in uasicircularorbits. Class. Quantum Gravity **25**(16), 165003 (2008)
4. M.A. Abramowicz, W. Kluzniak, A precise determination of black hole spin in GRO J1655–40. A & A **374**, L19–L20 (2001)
5. T.E. Strohmayer, Discovery of a 450 HZ quasi-periodic oscillation from the microquasar GRO J1655–40 with the Rossi X-ray timing explorer. Ap J **552**, L49–L53 (2001)
6. M. Gierlinski, A. Maciołek-Niedźwiecki, K. Ebisawa, Application of a relativistic accretion disc model to X-ray spectra of LMC X-1 and GRO J1655–40. MNRAS **325**, 1253–1265 (2001)
7. G. Faye, L. Blanchet, A. Buonanno, Higher-order spin effects in the dynamics of compact binaries. I. Equations of motion. Phys. Rev. D **74**(10), 104033 (2006)
8. E.E. Flanagan, T. Hinderer, Constraining neutron-star tidal Love numbers with gravitational-wave detectors. Phys. Rev. D **77**, 021502 (2008)
9. T. Hinderer et al., Tidal deformability of neutron stars with realistic equations ofstate and their gravitational wave signatures in binary inspiral. Phys. Rev. D **81**(12), 123016 (2010)
10. E. Poisson, Gravitational waves from inspiraling compact binaries: the uadrupolemomentterm. Phys. Rev. D **57**, 5287–5290 (1998)
11. K. Yagi, N. Yunes, I-Love-Q. ArXiv e-prints (2013), arXiv:1302.4499
12. K. Yagi, N. Yunes, *I-Love-Q Relations in Neutron Stars and Their Applications to Astrophysics Gravitational Waves and Fundamental Physics*, ArXiv e-prints (2013), arXiv:1303.1528
13. A. Maselli et al., I-Love-Q forever. ArXiv e-prints (2013), arXiv:1304.2052
14. P.C. Peters, J. Mathews, Gravitational radiation from point masses in a keplerian orbit. Phys. Rev. **131**, 435–440 (1963)
15. P.C. Peters, Gravitational radiation and the motion of two point masses. Phys. Rev. **136**, B1224–B1232 (1964)
16. R.M. O'Leary, B. Kocsis, A. Loeb, Gravitational waves from scattering of stellar-mass black holes in galactic nuclei. MNRAS **395**, 2127–2146 (2009)
17. L. Wen, On the eccentricity distribution of coalescing black hole binaries driven by the Kozai mechanism in globular clusters. ApJ **598**, 419–430 (2003)
18. W.E. East, F. Pretorius, Dynamical capture binary Neutron star mergers. ApJ **760**(L4), L4 (2012)
19. A. Gopakumar, B. R. Iyer, Second post-Newtonian gravitational wave polarizations for compact binaries in elliptical orbits. Phys. Rev. D **65**(8), 084011 (2002)

20. L. Blanchet, Innermost circular orbit of binary black holes at the third post-Newtonian approximation. Phys. Rev. D **65**(12), 124009 (2002)
21. M. Campanelli et al., Comparison of numerical and post-Newtonian waveforms for generic precessing black-hole binaries. Phys. Rev. D **79**(8), 084010 (2009)
22. I. Gradshteyn, I. Ryzhik, Table of integrals, series, and products academic. New York **19652**, 717 (1980)
23. T. Damour, B. R. Iyer, B. S. Sathyaprakash, Comparison of search templates for gravitational waves from binaryinspiral. Phys. Rev. D **63**(4), 044023 (2001)
24. T. Damour, B. R. Iyer, B. S. Sathyaprakash, Comparison of search templates for gravitational waves from binaryinspiral: 3.5PN update. Phys. Rev. D **66**(2), 027502 (2002)
25. K. G. Arun et al. Parameter estimation of inspiralling compact binaries using 3.5post-Newtonian gravitational wave phasing: the nonspinning case. Phys. Rev. D **71**(8), 084008 (2005)
26. A. Buonanno et al., Comparison of post-Newtonian templates for compact binary inspiralsignals in gravitational-wave detectors. Phys. Rev. D **80**, 084043 (2009)

Chapter 3
Gravitational Waves: Detection and Sources

3.1 Gravitational Wave Detectors

In Sect. 1.4, we have shown that a GW causes the stretching and shrinking of the proper distance between test particles. Therefore, GWs can be observed by measuring these changes in the proper distance. The focus of this work will be on the use of interferometers to measure the influences of GWs. These interferometers are configured to show destructive interference when no distortions are present, as shown Fig. 3.1a. When a GW passes such a detector, the configuration will no longer support the destructive interference and a bright spot can be seen, as shown in Fig. 3.1b. By measuring the intensity of this spot as well as its temporal variation, one will be able to observe the footprints of a GW.

3.1.1 Beam Pattern Functions

Interferometers are sensitive to the relative difference between two distances, the so-called *strain*. Suppose we have an interferometer with its arms pointing along the unit vectors u^i and v^i. One can show [2] that the strain, $h(t)$, is given by

$$\begin{aligned} h(t) &= \frac{1}{2}(h_{ij}u^i u^j - h_{ij}v^i v^j) \\ &= D^{ij}h_{ij}(t), \end{aligned} \tag{3.1}$$

where D^{ij} is referred to as the detector tensor and is given by

$$D^{ij} = \frac{1}{2}\left(u^i u^j - v^i v^j\right). \tag{3.2}$$

T.G.F. Li, *Extracting Physics from Gravitational Waves*,
Springer Theses, DOI 10.1007/978-3-319-19273-4_3

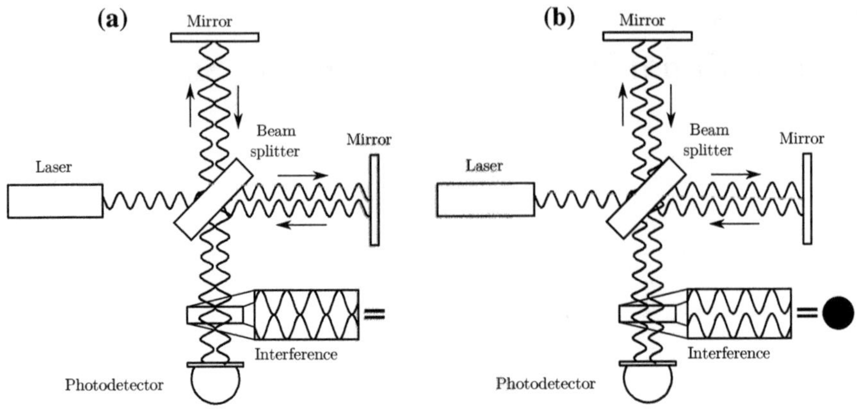

Fig. 3.1 Schematic overview of an interferometric detector [1]. Without a GW present, the interferometer is kept at a destructive interference configuration. A GW will change the relative distance between the arms and cause the interferometer output to deviate from the destructive interference. The intensity of the light and its temporal evolution encode the footprints of GWs. **a** No GW present, **b** GW present

Furthermore, as Eq. (3.1) is linear in h_+ and $h_\times$, we can also write

$$h(t) = F_+ h_+(t) + F_\times h_\times(t), \tag{3.3}$$

where $F_{+,\times}$ are called the *beam pattern functions*.

Suppose we have a detector with arms that are perpendicular to each other, one pointing in the x-direction and the other in the y-direction in a Cartesian coordinate system. This detector frame, denoted by (x, y, z), is generally different from the GW coordinate system, denoted by (x', y', z'), where the source is conveniently described. To account for such a difference, we first note that when the plus and cross polarisations are not equal in strength, we can rotate the coordinate system by an angle ψ around the z' axis so that the x' and y' axes coincide with the mayor and minor axis of the associated ellipse. In going from the GW frame to the detector frame, we can rotate the GW frame by an angle θ around the x' axis and an angle ϕ around the z' axis, where the angles (θ, ϕ) denote the direction of propagation of the GW in the detector frame. Applying these three rotations, the beam pattern functions for a detector with perpendicular arms are given by

$$F_+^{(90°)} = \frac{1}{2}\left(1 + \cos^2\theta\right)\cos 2\phi \cos 2\psi - \cos\theta \sin 2\phi \sin 2\psi, \tag{3.4}$$

$$F_\times^{(90°)} = \frac{1}{2}\left(1 + \cos^2\theta\right)\cos 2\phi \sin 2\psi + \cos\theta \sin 2\phi \cos 2\psi. \tag{3.5}$$

The beam pattern functions are shown in Fig. 3.2. Suppose that the arms of the detector are at a 60° angle instead of a 90° angle. In this case, the beam pattern

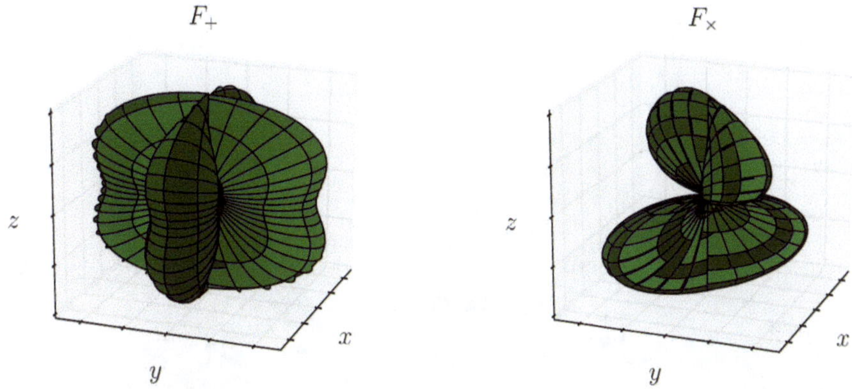

Fig. 3.2 Beam pattern functions $F_+(\theta, \phi, \psi)$ (*left*) and $F_\times(\theta, \phi, \psi)$ (*right*) for an interferometric detector with arms at a 90° angle

functions are given by

$$F_+^{(60°)} = \frac{\sqrt{3}}{2}\left[\frac{1}{2}(1+\cos^2\theta)\cos 2\phi\cos 2\psi - \cos\theta\sin 2\phi\sin 2\psi\right], \tag{3.6}$$

$$F_\times^{(60°)} = \frac{\sqrt{3}}{2}\left[\frac{1}{2}(1+\cos^2\theta)\cos 2\phi\sin 2\psi + \cos\theta\sin 2\phi\cos 2\psi\right]. \tag{3.7}$$

Compared to a detector with arms at a 90° angle, a detector with a 60° opening angle is less sensitive to GWs. However, with a 60° opening angle, one is able to fit three detectors forming an equilateral triangle in the same location. Especially when the detectors are to be underground, this will significantly save on costs for e.g. digging tunnels.

3.1.2 Ground-Based Detectors

At the moment, there are several ground-based interferometric gravitational waves detectors, either operational or in an upgrade phase. Only GEO (600 m arms) in Germany [3] is currently taking data. However, the biggest of these detectors, LIGO (two interferometers, 4 km arm length) in the US [4] and Virgo (3 km arm length) in Italy [5] are currently being upgraded from the initial/enhanced configurations (collectively referred to as 1st generation) to the so-called advanced configuration (2nd generation). An aerial view of the LIGO detector is shown in Fig. 3.3. This round of major upgrades is expected to complete in 2014 and will increase their sensitivity significantly. A Japanese detector named KAGRA [6] is currently under

Fig. 3.3 Aerial view of the LIGO detector in Hanford. Photo courtesy of the LIGO Laboratory

construction, and is expected to be operational in 2018. Finally, India aims to have an advanced detector named LIGO-India running by 2020 [7].

Plans are also made for detectors beyond the 2nd generation. The most advanced proposal is the so-called Einstein Telescope (ET), a European initiative for a 3rd generation detector. A conceptual design study has been published proposing three co-located, underground detectors, each with 10 km arms with a 60° opening angle [1]. Such a design admits even better sensitivity compared to 2nd generation detectors, up to a factor of ten (see Fig. 3.4).

3.2 Detection of Gravitational Waves

So far we have only considered the effect of a GW on a ring of test particles. In an actual detector, we will not only see the effect of GWs on the test masses (mirrors in case of an interferometer), but also noise contributions such as human traffic. If we denote the noise contribution as $n(t)$ and the gravitational wave induced strain as $h(t)$, then the total output of the detector is given by

$$s(t) = n(t) + h(t). \tag{3.8}$$

Therefore, not only will we have to understand the effects of GWs, it is also important to understand the noise in order to find the GW footprint in the detector output. We will consider each of these contributions in turn.

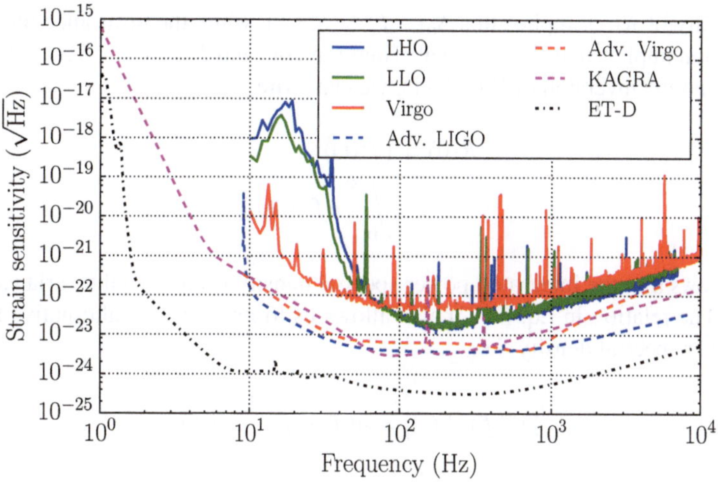

Fig. 3.4 Detector sensitivity for various detectors in terms of the strain sensitivity, $\sqrt{S_n(f)}$. The *solid curves* represent the strain sensitivity of LIGO in Hanford (LHO, *blue*), LIGO in Livingston (LLO, *green*) and Virgo in Pisa (*red*) on their last joint science run, before the upgrade to their advanced configurations. The *dashed curves* show the design sensitivity for the second generation detectors. Advanced LIGO (*blue*), Advanced Virgo (*red*), KAGRA (*magenta*). The *dashed-dotted curve* shows the design sensitivity labelled ET-D for the third generation detector Einstein Telescope (*black*)

3.2.1 Characterising the Noise

Suppose we can describe the noise $n(t)$ by a Gaussian stochastic process [8]. A Gaussian stochastic process is uniquely characterised by its average value and its auto-correlation. Without loss of generality, we can set the average value to be

$$\langle n(t)\rangle = 0, \tag{3.9}$$

where the brackets denote the ensemble average. In practice, one only has access to a single realisation of the detector. Therefore, we assume ergodicity and replace the ensemble average with the time average. The auto-correlation, defined as

$$R(\tau) \equiv \langle n(t+\tau)n(t)\rangle, \tag{3.10}$$

typically decays to zero when $|\tau| \to \infty$, i.e. the knowledge of the noise at time t reveals little information about the noise at a time $t + \tau$. We can now define the *one-sided power spectral density* as

$$\frac{1}{2}S_n(f) \equiv \int_{-\infty}^{\infty} d\tau\; R(\tau)e^{\imath 2\pi f\tau}. \tag{3.11}$$

The reality of $R(\tau)$ implies that $S_n(-f) = S_n^*(f)$, while the invariance under time translation implies that $S_n(f)$ is real. Therefore, one has $S_n(-f) = S_n(f)$. With Eq. (3.10) and the inversion of Eq. (3.11), we can write

$$R(0) = \langle n^2(t) \rangle \\ = \int_0^\infty df \; S_n(f). \tag{3.12}$$

This shows that the integrated power spectral density is equal to the variance of the noise. The relation in Eq. (3.12) also allows us to introduce an alternative form of the power spectral density, i.e.

$$\langle \tilde{n}^*(f)\tilde{n}(f') \rangle = \delta(f - f')\frac{1}{2}S_n(f), \tag{3.13}$$

where a tilde denotes the Fourier transform and a star the complex conjugate. The (expected) power spectral densities for various detectors are shown in Fig. 3.4.

3.2.2 *Matched Filtering*

In the last subsection, we looked at characterising the noise. Naively, one might expect that the signal $h(t)$ can only be detected if $|n(t)| < |h(t)|$. However, for ground-based GW detectors, we are in the regime where $|n(t)| \gg |h(t)|$. It turns out that, provided one has information on $h(t)$, we can still find the GW within the data even if the noise amplitude is bigger than the GW amplitude.

Suppose we know the shape of $h(t)$. One can compute the average of the product of $s(t)$ with $h(t)$ over a period of T

$$\frac{1}{T}\int_0^T dt \; s(t)h(t) = \frac{1}{T}\int_0^T dt \; h^2(t) + \frac{1}{T}\int_0^T dt \; n(t)h(t). \tag{3.14}$$

The integral in the first term on the RHS is positive definite and thus grows as T. In terms of the order of magnitude, we have

$$\frac{1}{T}\int_0^T dt \; h^2(t) \sim h_0^2, \tag{3.15}$$

where h_0 is the characteristic amplitude of the oscillating function $h(t)$. On the other hand, in the integral in the second term on the RHS, $n(t)$ and $h(t)$ are uncorrelated and the integral thus grows like $T^{1/2}$, similar to a random walk. Therefore, we have

$$\frac{1}{T}\int_0^T dt \; n(t)h(t) \sim \sqrt{\frac{\tau_0}{T}}\, n_0 h_0, \tag{3.16}$$

where n_0 is the characteristic amplitude of the noise $n(t)$ and τ_0 is a characteristic timescale, e.g. the period of $h(t)$. Therefore, it is not necessary to have $|n(t)| < |h(t)|$. Instead, it suffices to have

$$h_0 > \sqrt{\frac{\tau_0}{T}}\, n_0. \tag{3.17}$$

Thus, if we integrate the signal over long period compared to τ_0, we can average the noise to zero and detect the signal even when $|n(t)| > |h(t)|$. To exemplify, a binary system has a characteristic frequency of $f \sim 100\,\text{Hz}$, or $\tau_0 \sim 10^{-2}\,\text{s}$ when it is detectable by ground-based detectors. Such signals can be observed in the order of minutes, say $T \sim 100\,\text{s}$. Therefore, we have $\sqrt{\tau_0/T} \sim 10^{-2}$, meaning that the signal can still be detected if the signal amplitude is factor of 100 smaller than the noise amplitude. For a continuous signal, such as a millisecond pulsar, this factor can be even larger. Suppose we have $\tau_0 \sim 10^{-3}\,\text{s}$, but we can observe the signal for a period of a year, $T \sim 10^7\,\text{s}$. Using these numbers, we have $\sqrt{\tau_0/T} \sim 10^{-5}$, meaning we are sensitive to signals with an amplitude of a factor of 100,000 smaller than the noise amplitude.

We can optimise the above by the application of a technique called *matched filtering*. We define

$$\hat{s} = \int_{-\infty}^{\infty} dt\; s(t)K(t), \tag{3.18}$$

where $K(t)$ is called a filter. Furthermore, we define a signal-to-noise ratio (SNR), $\rho = S/N$, where S is the expectation value of $\hat{s}$ if a GW signal is present and N is the root mean square value of $\hat{s}$ when no GW signal is present. The idea behind matched filtering is to find the filter that maximises the SNR. Since $\langle n(t)\rangle = 0$, we have

$$\begin{aligned} S &= \int_{-\infty}^{\infty} dt\; \langle s(t)\rangle K(t) \\ &= \int_{-\infty}^{\infty} dt\; h(t)K(t) \\ &= \int_{-\infty}^{\infty} df\; \tilde{h}(f)\tilde{K}^*(f). \end{aligned} \tag{3.19}$$

Furthermore, for $h(t) = 0$ we have

$$\begin{aligned} N^2 &= \langle \hat{s}^2\rangle - \langle \hat{s}\rangle^2 \\ &= \langle \hat{s}^2\rangle \end{aligned}$$

$$= \int_{-\infty}^{\infty} \int_{-\infty}^{\infty} dt\, dt'\; K(t)K(t')\langle n(t)n(t')\rangle$$
$$= \frac{1}{2} \int_{-\infty}^{\infty} df\; S_n(f) \left|\tilde{K}(f)\right|^2, \tag{3.20}$$

where we have used the definition of the power spectral density in Eq. (3.11). Subsequently, the SNR is given by

$$\rho = \frac{\int_{-\infty}^{\infty} df\; \tilde{h}(f)\tilde{K}^*(f)}{\sqrt{\int_{-\infty}^{\infty} df\; \frac{1}{2} S_n(f) \left|\tilde{K}(f)\right|^2}}. \tag{3.21}$$

This expression can be simplified by introducing an inner product between $a(t)$ and $b(t)$

$$(a|b) = \Re \int_{-\infty}^{\infty} df\; \frac{\tilde{a}^*(f)\tilde{b}(f)}{\frac{1}{2} S_n(f)}. \tag{3.22}$$

With this definition, we can rewrite Eq. (3.21) as

$$\rho = \frac{(u|h)}{\sqrt{(u|u)}}, \tag{3.23}$$

where $\tilde{u}(f)$ is given by

$$\tilde{u}(f) = \frac{1}{2} S_n(f) \tilde{K}(f). \tag{3.24}$$

From this geometrical representation, we can deduce that ρ is maximised when u points in the same direction as h. The optimal filter is therefore given by the so-called *Wiener filter*

$$\tilde{K}(f) \propto \frac{\tilde{h}(f)}{S_n(f)}. \tag{3.25}$$

Furthermore, the SNR for the Wiener filter, also referred to as the *optimal* SNR, is given by

$$\rho = \sqrt{(h|h)}. \tag{3.26}$$

From the above results, we can conclude that the optimal filter depends on the signal searched for, and the properties of the noise. It is therefore critical to gain a good understanding of both aspects.

Assuming that the noise is uncorrelated between detectors, we can also define the optimal SNR for a *network* of detectors by adding the SNR of individual detectors in quadrature. In other words, the network SNR is given by

$$\rho_{\text{net}}^2 = \sum_i \rho_i^2, \tag{3.27}$$

where the index i runs over individual detectors.

In the derivation of the matched filtering, we have assumed that the shape of the true signal and the characteristics of the noise are known. For GW analyses, however, several issues are present. Firstly, the characterisation of the noise is not trivial. For example, the noise might not be stationary and Gaussian. Moreover, there is an inherent ambiguity whether a GW signal is being included in the noise characterisation. Secondly, one rarely has access to the exact same GW signal in the filter. For example, for most realistic GW sources, the waveform is only known in the form of an approximation (cf. the PN formalism in Chap. 2). Furthermore, as the GW signal depends on many parameters, the discretisation of the collection of filters used can cause an SNR lower than the SNR given by the exact Wiener filter. We will further discuss the issues concerning the interpretation of GW signals in Chap. 4.

3.3 Sources of Gravitational Waves

Although every accelerating non-symmetric system emits GWs, only the ones that are able to perturb the metric enough can be detected with current technology. Naturally, we are looking for those systems that are massive, compact and/or violent enough to induce a sufficiently strong gravitational field. It turns out that such sources are only of astrophysical origin, such as black holes (BHs) and neutron stars (NSs) or violent events like supernovae and gamma-ray bursts. In this section, we will look at some of the most promising sources for detection by ground-based detectors. The study of the sources of GWs is an extensive field and citing only a few references will not do justice to the large body of work performed by numerous authors. Therefore, unless explicitly stated, references and/or proofs to relevant statements can be found in Ref. [2] and references therein.

3.3.1 Compact Binary Coalescences

Compact binary coalescence (CBC) is a class of sources in which two compact objects, either a NS or a BH, orbit around each other. CBC sources are further subdivided into binary neutron stars (BNS), black hole neutron star binary (BHNS) and binary black hole (BBH). The emission of GWs carries energy and angular momentum away from the system, causing the two objects to spiral towards each other. The most famous CBC system is perhaps the Hulse-Taylor binary pulsar [9].

The orbital evolution of the Hulse-Taylor binary pulsar provided the first indirect evidence of the emission of GWs. GWs from CBC sources can be divided into three stages, the inspiral, the merger and the ringdown phase.

The *inspiral* phase is when the two object are spiralling towards each other, while the system loses orbital energy and angular momentum. As a result, the GW amplitude and frequency slowly increase as a function of time, the so-called *chirp*. Sources can remain in the inspiral phase for hundreds of millions of years, but they can only be observed by ground-based detectors towards the end of the inspiral phase. During the inspiral phase, the gravitational fields and velocities are still relatively weak and the GWs emitted can be found through approximation methods such as the PN formalism in Chap. 2.

The *merger* phase is a short-lived phase that follows the inspiral phase and occurs when the two objects are so close to each other that they start to merge into a single object. The gravitational fields are now very strong and the GW emission can only be computed by considering the full EFE numerically. As it turns out, the merger phase can cause a GW luminosity that exceeds the electromagnetic luminosity of the entire Universe.

Once the two objects have merged, they enter the *ringdown* phase. During this phase, the newly formed object tries to reach a quiescent state by radiating away its deformations from the merger phase. The GWs in the ringdown phase can be computed by means of perturbation theory. The resulting GWs consist of superpositions of damped sinusoids, known as quasi-normal modes.

3.3.2 Continuous Wave Sources

Continuous wave (CW) sources are those that emit GWs with roughly constant frequency and amplitude compared to the observation time. The prime candidates to emit such signals are rapidly rotating, non-axisymmetric NSs. Such deformations can arise due to, for example, strain build-up in the crust or in the core, or accretion of matter.

The GW signal from CW sources is relatively simple due to their slow variation of amplitude and frequency. Although the GW amplitude is generally weaker compared to CBC sources, a longer integration time means that CW sources may also achieve detectable SNRs. Finally, the analysis of such sources can be simplified if some of its characteristics can be determined through means other than GWs, e.g. rotational frequency of a NS through the observation of a pulsar.

3.3.3 Burst Sources

Burst sources are associated with astronomical transient phenomena, such as supernovae, gamma-ray bursts or instabilities in NSs. Supernovae occur when massive objects collapse under the influence of gravity. In the process that follows, NSs or

BHs are formed, and GWs are emitted through dynamical processes. On the other hand, GRBs are flashes of gamma rays, for which the progenitors are still uncertain. Possible progenitors are supernova-like events (causing long duration soft spectrum GRBs, also called long soft GRBs) as well as the merger of compact objects (causing short hard GRBs).

These sources are generally difficult to model because of the complicated physics associated with such phenomena. Therefore, the search for GWs from such sources often happens with unmodelled filters, where unmodelled means that there is no astrophysical model associated to the GW. An example of such a filter is a sine-Gaussian signal.

3.3.4 Stochastic Background

The stochastic background comes from the superposition of numerous unresolved GW sources. The stochastic background is divided into two classes: the primordial background and the astrophysical background. The primordial background consists of radiation originating from the early Universe, such as the Big Bang. The astrophysical background comes from the GWGW radiation from astrophysical sources such as CBC systems or cosmic-string cusps and kinks.

The detection of a stochastic background is somewhat different compared to the sources described above. Since random radiation is indistinguishable from instrumental noise, at least for short observation times, it cannot be detected with a single detector. Instead, the output of several detectors are combined to calculate the cross-correlation between detectors. An excess in the cross-correlation can then be an indication of a stochastic background.

References

1. Einstein Telescope Science Team. Einstein gravitational wave Telescope conceptual design study. Technical report ET-0106C-10, 2011
2. B. Sathyaprakash, B.F. Schutz, Physics, astrophysics and cosmology with gravitational waves. Living Rev. Relat. **12**(2) (2009)
3. H Grote, The LIGO Scientific Collaboration, The GEO 600 status. Class. Quantum Gravity **27**(8), 084003 (2010)
4. B. P. Abbott et al., LIGO: the laser interferometer gravitational-wave observatory. Rep. Progr. Phys. **72**(7) (2009)
5. T. Accadia et al., Status of the Virgo project. Class. Quantum Gravity **28**(11), 114002 (2011)
6. K. Somiya, Detector configuration of KAGRA-the Japanese cryogenic gravitational wave detector. Class. Quantum Gravity **29**(12), 124007 (2012)
7. B. Iyer et al., LIGO-India, Proposal of the consortium for indian initiative in gravitational-wave observations (IndIGO). Technical report, 2011
8. J.L. Doob, J.L. Doob, *Stochastic Processes*, vol. 7.2 (Wiley, New York, 1953)
9. R.A. Hulse, J.H. Taylor, Discovery of a pulsar in a binary system. Astrophys. J. Lett. **195**, L51–L53 (1975)

Chapter 4
Bayesian Inference

In this chapter, we introduce the field of Bayesian inference. Named after Reverend Bayes, Bayesian inference aims to give a consistent mathematical framework to *inductive logic*. Inductive logic, also referred to as *plausible reasoning*, is a subfield of formal logic that tries to evaluate logically certain statements. This should be viewed in contrast to *deductive logic*, which attempts to draw conclusions from logically certain statements.

After introducing some elementary concepts of inductive logic, we will show how Bayesian inference provides a mathematically consistent framework to two commonly required tasks within data analysis: parameter estimation and hypothesis testing.

4.1 Concepts of Inductive Logic

4.1.1 Definitions and Notations of Logic

Propositions
The central concept in the field of formal logic is a *proposition*. A proposition is defined as

> a statement in which a subject and a predicate are combined to assert its (un)truth.

Thus, a proposition can be the statement "hypothesis H is correct" or "quantity x has a value of 3". Proposition are usually denoted by an upper case roman letter, e.g. A and B.

Boolean Algebra
Operations on, or combinations of propositions can be conveniently written according to the rules of *Boolean* algebra. These operations or combinations are the following.

Conjunction The conjunction, also called the logical product, of A and B is denoted by

T.G.F. Li, *Extracting Physics from Gravitational Waves*,
Springer Theses, DOI 10.1007/978-3-319-19273-4_4

$$AB \quad \text{or} \quad A \wedge B, \tag{4.1}$$

and is associated to the proposition "both A and B are true". The conjunction operator is commutative.

Disjunction The disjunction, sometimes referred to as the logical sum, of A and B is denoted by

$$A + B \quad \text{or} \quad A \vee B, \tag{4.2}$$

and is associated to the proposition "at least one of A and B is true". Similarly to the conjunction, the disjunction is commutative.

Denial The denial, or negation, of a proposition A states that "A is false" and is denoted by

$$\bar{A} \quad \text{or} \quad \neg A. \tag{4.3}$$

We can further assert that the denial of the denial of A equals A again.

Implication The implication relates two propositions conditionally. The implication of A and B is defined to be "if A is true then B is true" and is written as

$$A \rightarrow B \quad \text{or} \quad A \Rightarrow B. \tag{4.4}$$

Bi-Implication An even stronger relational statement between two propositions can be made in the form of a bi-implication. The bi-implication of A and B represents the proposition "B is true if and only if A is true" and is denoted by

$$A \leftrightarrow B \quad \text{or} \quad A \Leftrightarrow B. \tag{4.5}$$

In the language of Boolean algebra, when two propositions are related by a bi-implication they have the same *truth value*. From a logical point of view, they are hence considered to be equivalent propositions and the bi-implication can thus also be written as

$$A = B. \tag{4.6}$$

Similarly, we also have the "equal by definition", denoted by

$$A \equiv B. \tag{4.7}$$

4.1.2 Foundations of Inductive Logic

The study of inductive logic is in an extension of the Aristotelian logic. Instead of assigning either true or false to a proposition, one assigns a degree of certainty to a proposition. This degree of certainty, also referred to as *plausibility*, is always dependent on what we consider to be known. For example, if we know that the sky is cloudy, the plausibility of rain to come soon is higher compared to a sunny sky. Therefore, we expand the notation of Boolean algebra with the conditional plausibility "A is true given that B is true" or simply "A given B". The *conditional plausibility* is denoted by

$$A|B. \tag{4.8}$$

Cox' Postulates and Axioms

In search of mathematical rules for plausible reasoning, Cox wrote down a set of postulates which such a theory should obey. Following Jaynes [1], these can be stated as a set of desiderata:

- **Desideratum 1** Plausibilities are represented by real numbers.
- **Desideratum 2** The theory should have a qualitative correspondence with common sense.
- **Desideratum 3** The theory should be consistent. If a conclusion can be reasoned out in more than one way, then every possible way must lead to the same result.

However, these desiderata lack the mathematical rigor from which the rules of plausible reasoning can be derived. Instead, the desiderata can be augmented with mathematical axioms to produce the necessary proofs. Let the plausibility of a proposition A be denoted by $w(A)$. Cox' axioms can then be stated as [2]

- **Axiom 1** Degree of belief can be ordered; if $w(A)$ is greater than $w(B)$, and $w(B)$ is greater than $w(C)$, then $w(A)$ is greater than $w(C)$. Consequently, plausibility can be mapped onto real numbers.
- **Axiom 2** The plausibility of A and its negation $\bar{A}$ are related, i.e. there is a function f such that

$$w(A) = f\left[w(\bar{A})\right]. \tag{4.9}$$

- **Axiom 3** The plausibility of a conjunction AB is related to the plausibility of the conditional proposition $A|B$ and the plausibility of B alone, i.e. there is a function g such that

$$w(AB) = g\left[w(A|B), w(B)\right]. \tag{4.10}$$

4.1.3 Quantitative Rules for Plausible Reasoning

Cox showed that under these axioms, consistent mathematical rules exist to describe the field of inductive logic. Firstly, the plausibility w can be mapped onto probabilities p, if one assumes that w ranges from 0 for impossibility to 1 for certainty (another choice is to have w range from 1 for certainty to infinity for impossibility). Therefore, we will use the terms plausibility and probability interchangeably. It turns out that the rules of plausible reasoning coincide with the laws of probability, given by the *sum rule*,

$$p(A|I) + p(\bar{A}|I) = 1, \tag{4.11}$$

and the *product rule*,

$$p(A, B|I) = p(A|B, I)P(B|I), \tag{4.12}$$

where the proposition I represents all the relevant background information that one is aware of before the assignment of plausibilities. Many other results can be derived from these two rules. Amongst the most useful ones for data analysis are *Bayes' theorem*,

$$p(A|B, I) = \frac{p(B|A, I)p(A|I)}{p(B|I)}, \tag{4.13}$$

and the *marginalisation rule*,

$$p(A|I) = p(A, B|I) + p(A, \bar{B}|I). \tag{4.14}$$

The marginalisation rule can be generalised in several ways. Firstly, assume we have a set of propositions $\{B_k\}$ that form a mutually exclusive,

$$p(B_k|B_l, I) = p(B_k|I) \quad \text{for } k \neq l, \tag{4.15}$$

and exhaustive set of propositions,

$$\sum_k p(B_k|I) = 1. \tag{4.16}$$

In this case, the marginalisation rule can be written as

$$p(A|I) = \sum_k p(A, B_k|I). \tag{4.17}$$

We can go a step further and generalise the marginalisation rule to the continuum limit. But before we can do so, we need to explain what is meant by the continuum limit of a proposition. Take a proposition of the form "a continuous variable x has a value of α". We can write the probability of this proposition in the form $p(x = \alpha)$. However, since within the limits of a continuous variable there exist an arbitrary large number of propositions, we fail to assign a useful meaning to the probability of $x = \alpha$. So instead, we can only define the probability associated to a continuous variable in a finite interval, i.e.

$$p(x_1 \leq x \leq x_2|I) = \int_{x_1}^{x_2} \mathrm{pdf}(x)dx, \tag{4.18}$$

where pdf represents the *probability density function*. For a continuous variable, the property of exhaustiveness is written as

$$\int_{x_{\min}}^{x_{\max}} \mathrm{pdf}(x)dx = 1, \tag{4.19}$$

where the integration limits include all possible values of x. The marginalisation rule for continuous variables becomes

$$p(A) = \int_{x_{\min}}^{x_{\max}} \mathrm{pdf}(A, x)dx. \tag{4.20}$$

At this point, it should be noted that the probability $p(B)$ is dimensionless, whereas the probability density function $\mathrm{pdf}(x)$ has the inverse dimension of x. However, in the literature, the symbols p or P are used for both the probability and the probability density. The reader is expected to distil the meaning from the context.

4.2 Parameter Estimation

We can now use the rules in Sect. 4.1.3 to infer the value of a parameter given the observed data, commonly referred to as *parameter estimation*. Examples of parameter estimation include the inference of the mass of a black hole through the measurement of its gravitational waves, or the measurement of the Hubble constant by observing a set of standard candles.

Consider a model H that provides a description of the data d provided the parameter θ is known. In this case, model H is referred to as *generative*. The plausibility of the data d is then governed by the *likelihood* $p(d|\theta, H, I)$, where I represents the state of knowledge prior to the experiment. In most experiments, we have enough knowledge about the model H to assign the likelihood function. For example, given that we know we are flipping a biased coin that gives, on average, 7 heads out of every 10 flips, we can assign the likelihood of observing m heads from n throws.

However, in most experiments we are only exposed to the data d and are interested in the so-called *posterior* probability of the parameter θ, $p(\theta|d, H, I)$. The aim of this section is to show how to go from the likelihood to the posterior, whilst being consistent with the rules of plausible reasoning.

4.2.1 From Bayes' Theorem

Suppose we have a generative model H, and can therefore assign the likelihood of the data d. The posterior is related to the likelihood through Bayes' theorem in Eq. (4.13)

$$p(\theta|d, H, I) = \frac{p(d|\theta, H, I)p(\theta|H, I)}{p(d|H, I)}. \tag{4.21}$$

Two factors in this equation have yet to be given a name. Firstly, we have the *prior* $p(\theta|H, I)$ which represent the state of knowledge of the parameter θ before the experiment is conducted. This could be the interval in which the parameter is expected or the distribution of values it is expected to have. Secondly, we have the *evidence* $p(d|H, I)$. This quantity, as it does not depend on the parameter of interest θ, will be ignored for the time being and Eq. (4.21) will be turned into the proportional statement

$$p(\theta|d, H, I) \propto p(d|\theta, H, I)p(\theta|H, I). \tag{4.22}$$

We will revisit the evidence in Sect. 4.3 where we show that this factor plays a crucial role in the case that the generative model is the object of interest.

Equation (4.22) shows that the inference on the posterior is given by the likelihood weighted by the prior. This seems to be in line with desideratum 3, set out in Sect. 4.1.2, as common sense would dictate that one forms its conclusions based on the information that one has before the experiment and the experimental data that one obtains.

4.2.2 Characterising the Posterior

Once we have calculated the posterior, we have acquired all the information about the variables given the observed data. Suppose we are only interested in a single variable. We can plot the posterior as a function of this variable. From this plot, we can infer the values that are more likely than others, simply by looking at the height of the plot. However, this might not always be the desired representation. For example, plots in dimensions higher than three are difficult to interpret. Furthermore, one might want to summarise the posterior with only a few numbers, instead of a plot.

Marginalisation

Suppose we have performed an experiment governed by N parameters θ_i where $i = 1, 2, \ldots, N$, which we try to infer. Using Bayes' theorem, we obtain the so-called *joint posterior*

$$p(\theta_1, \ldots, \theta_N | d, H, I). \tag{4.23}$$

But suppose that we only want to show the posterior for variable θ_1, given by $p(\theta_1 | d, H, I)$. We can achieve this through the marginalisation rule given by Eq. (4.20). Thus, we can obtain the desired posterior through

$$p(\theta_1 | d, H, I) = \int_{\theta_2^{\min}}^{\theta_2^{\max}} \cdots \int_{\theta_N^{\min}}^{\theta_N^{\max}} p(\theta_1, \ldots, \theta_N | d, H, I)\, d\theta_2 \ldots d\theta_N. \tag{4.24}$$

Therefore, one way to present a multidimensional posterior is to draw one or two dimensional marginalised posterior functions.

Mean, Standard Deviation

Sometimes it can be more convenient to summarise the (joint) posterior with a few numbers. For example, instead of plotting the full posterior, we want to convey the expected value and the associated uncertainty. In case of a one dimensional posterior, we can calculate the expectation value, also known as the mean, through

$$\begin{aligned} \mu &= E\left[\theta\right] \\ &= \int_{\theta^{\min}}^{\theta^{\max}} \theta\, p(\theta | d, H, I)\, d\theta. \end{aligned} \tag{4.25}$$

A measure of the uncertainty can be represented by the second central moment, also referred to as the variance, which is given by

$$\begin{aligned} \sigma^2 &= E\left[(\theta - \mu)^2\right] \\ &= \int_{\theta^{\min}}^{\theta^{\max}} (\theta - \mu)^2\, p(\theta | d, H, I)\, d\theta. \end{aligned} \tag{4.26}$$

The multi-dimensional equivalents of the mean and the variance can also be defined. Suppose an experiment has N variables, θ_i where $i = 1, 2, \ldots, N$. The means of these variables are then given by

$$\begin{aligned} \mu_i &= E\left[\theta_i\right] \\ &= \int_{\theta_1^{\min}}^{\theta_1^{\max}} \cdots \int_{\theta_N^{\min}}^{\theta_N^{\max}} \theta_i\, p(\theta_1, \ldots, \theta_N | d, H, I)\, d\theta_1 \ldots d\theta_N. \end{aligned} \tag{4.27}$$

Similarly, a higher dimensional variance-like object called the *covariance matrix* can be defined as

$$\Sigma_{ij} \equiv E\left[(\theta_i - \mu_i)\left(\theta_j - \mu_j\right)\right] \\ = \int_{\theta_1^{\min}}^{\theta_1^{\max}} \cdots \int_{\theta_N^{\min}}^{\theta_N^{\max}} (\theta_i - \mu_i)\left(\theta_j - \mu_j\right) p(\theta_1, \ldots, \theta_N | d, H, I)\, d\theta_1 \ldots d\theta_N. \tag{4.28}$$

The diagonal elements give the variances of the variables, whereas the off-diagonal elements represent the correlations between the variables.

Confidence Interval

The variance is a good measure of the uncertainty if the posterior is symmetric. However, if the posterior is asymmetric, then the variance fails to report the direction to which the posterior is skewed. Of course, one can provide the skewness (3rd central moment) alongside the mean and the variance. In this case, a better representation of the posterior is given by the *confidence interval*. The confidence interval is defined to be the smallest limits within which a fraction γ of the posterior is contained. Formally, this is given by

$$\gamma = \int_{\theta^{\text{lo}}}^{\theta^{\text{hi}}} p(\theta | d, H, I) d\theta, \tag{4.29}$$

where the distance between θ^{hi} and θ^{lo} is minimised. In most literature, γ is either chosen to be 0.68 or 0.95, which are the fractions of the total probabilities given by the 1 and 2 standard deviation intervals in a Gaussian distribution, which follows from the central limit theorem [2].

The multidimensional confidence interval can also be defined as an area in two dimensions, volume in three dimensions and hypervolume in even higher dimensions. However, the information quickly becomes difficult to summarise, something we set out to do by defining the confidence interval. More useful would be to show the confidence interval of the marginalised posteriors. For θ_1, this is given by

$$\gamma_{\theta_1} = \int_{\theta_1^{\text{lo}}}^{\theta_1^{\text{hi}}} p(\theta_1 | d, H, I) d\theta_1 \\ = \int_{\theta_1^{\text{lo}}}^{\theta_1^{\text{hi}}} \int_{\theta_2^{\min}}^{\theta_2^{\max}} \cdots \int_{\theta_N^{\min}}^{\theta_N^{\max}} p(\theta_1, \ldots, \theta_N | d, H, I) d\theta_1 \ldots d\theta_N. \tag{4.30}$$

Multimodality

In the case of multimodal posteriors, providing a summary in just a few numerical values can be troublesome. The mean value could occur in a location with low probability. The variance and the confidence interval can span multiple modes, thus suggesting that the data are expected with a larger spread than actually is expected by looking at the individual modes.

Unfortunately, there is no generic way of summarising multimodal distributions. In cases where there are only a few modes, one could try to characterise each single mode through the mean, standard deviation or confidence interval. However, if the posterior exhibits more complicated structures, the most honest thing to do is to show the full or marginalised posterior where possible.

The practice of estimating parameters is intimately connected to all aspects of experimental science. GWs from CBC systems possess the unique property of carrying information about the distance from the source to the observer. Being able to infer this distance allows us to map the dynamics and the content of the Universe. In part III, we will use the concepts of parameter estimation to investigate the power of using GW sources to infer the cosmological parameters that dictate the large scale properties of our Universe.

4.3 Hypothesis Testing

Estimating parameters is useful when the generative model is known, but this not always the case in an experiment. For example, suppose we measure a process of which we do not know whether its output follows a normal distribution or a Lorentzian distribution. If we want to compare generative models to each other, we have to perform *hypothesis testing*, also known as *model comparison* or *model selection.* In order to compare hypothesis X to hypothesis Y, we must calculate the posterior probabilities $p(X|d, I)$ and $p(Y|d, I)$ and compare the two.

4.3.1 From Bayes' Theorem

Just as with parameter estimation, we want to calculate the posterior probability through quantities that are easier to compute. To do this, we again turn to Bayes' theorem and write

$$p(X|d, I) = \frac{p(d|X, I)p(X|I)}{p(d|I)}. \tag{4.31}$$

In order to compare model X with model Y, we take the ratio of the posterior probabilities and compute the so-called *odds ratio*,

$$
\begin{aligned}
O_Y^X &\equiv \frac{p(X|d, I)}{p(Y|d, I)} \\
&= \frac{p(d|X, I)}{p(d|Y, I)} \frac{p(X|I)}{p(Y|I)},
\end{aligned} \tag{4.32}
$$

where the factors of $p(d|I)$ have cancelled out. The factor $p(X|I)/p(Y|I)$ is called the *prior odds* and expresses our relative initial belief in the hypotheses. The factor $p(d|X, I)/p(d|Y, I)$ is the ratio of evidences, commonly known as the *Bayes factor*,

$$
B_Y^X = \frac{p(d|X, I)}{p(d|Y, I)}. \tag{4.33}
$$

Thus, in order to do hypothesis testing, one needs to calculate the evidence for propositions X and Y, and assign the prior odds between the same propositions.

We have seen the evidence before in Eq. (4.21). We can rewrite Bayes' theorem in Eq. (4.21) as

$$
p(\theta|d, H, I)p(d|H, I) = p(d|\theta, H, I)p(\theta|H, I), \tag{4.34}
$$

and marginalise both sides over θ, i.e.

$$
\int p(\theta|d, H, I)p(d|H, I)d\theta = \int p(d|\theta, H, I)p(\theta|H, I)d\theta. \tag{4.35}
$$

But as $p(d|H, I)$ does not depend on θ and the posterior is normalised by definition, we can evaluate the LHS to be

$$
\begin{aligned}
\int p(\theta|d, H, I)p(d|H, I)d\theta &= p(d|H, I) \int p(\theta|d, H, I)d\theta \\
&= p(d|H, I).
\end{aligned} \tag{4.36}
$$

Therefore, the evidence is given by

$$
p(d|H, I) = \int p(d|\theta, H, I)p(\theta|H, I)d\theta, \tag{4.37}
$$

and can be interpreted as the fully marginalised, prior weighted likelihood, also referred to as the *marginal likelihood*. As we expect from common sense, the evidence does not depend on the value of θ.

The odds ratio thus tells us which of the propositions is favoured compared to another. If $O_Y^X > 1$, the data tell us that proposition X is more likely, if $O_Y^X = 1$ the propositions are equally likely, and if $O_Y^X < 1$, proposition Y is favoured over X. But the odds ratio also tells us how much a proposition is favoured over another through the magnitude of the odds ratio. If $O_Y^X = 5$, proposition is 5 times more likely compared to Y.

Up to now, we have shown how to compare two propositions. Can we calculate the absolute probability of a single proposition from Eq. (4.31)? We have seen that the evidence and the prior can be assigned. But what about the factor $p(d|I)$? Can we assign the probability of the data given the background information? The probability of the data can be written as

$$\begin{aligned} p(d|I) &= p(d, X|I) + p(d, \bar{X}|I) \\ &= p(d|X, I)p(X|I) + p(d|\bar{X}, I)p(\bar{X}|I) \end{aligned} \tag{4.38}$$

Although the prior probabilities can be related through

$$p(X|I) + p(\bar{X}|I) = 1, \tag{4.39}$$

assigning an evidence to $\bar{X}$ is, in most cases, impossible, as one needs a generative model instead of the denial of a single proposition. To illustrate this point, we can predict the signal of a GW assuming GR is correct, but we cannot predict the signal of a GW when we only know that GR is incorrect. Therefore, in order to perform inference on generative models, we need to compare one generative model to another.

4.3.2 Occam's Razor

Does the model which fits the data best always give the highest evidence? If that is the case, one expects that the generative model with the most freedom to have a higher evidence compared to a more restrictive model. However, we will show that this is not generally true within Bayesian inference.

The fit of the data with model X is represented by the evidence $p(d|X, I)$. However, from Eq. (4.32), we see that the Bayes' factor is altered by the prior odds to give the odds ratio. Thus, a model needs to be particularly favoured by the data if prior belief in this model is lower compared to the other model. Therefore, important to hypothesis testing is the assignment of the background information, i.e. what is known prior to the experiment.

But that is not all. From common sense, it seems to be unfair if the more complex model is always favoured compared to a more restrictive model. Suppose the generative model is described by a linear relationship. One would hope that a consistent analysis of the data shows that the generative model was indeed the linear relationship instead of a general polynomial that can describe the data equally well through the appropriate choice of parameters. We can show how Bayesian inference works under such conditions by an example.

Suppose we have a generative model X, which has no free parameter, and generative model Y, which does have a free parameter λ. Will the more complicated model, Y, be automatically favoured compared to X? To answer this question, we turn to the odds ratio

$$O_Y^X = \frac{p(d|X, I)}{p(d|Y, I)} \frac{p(X|I)}{p(Y|I)}. \tag{4.40}$$

The assignment of the evidence for X is straightforward, but the evidence of Y depends on λ through

$$\begin{aligned} p(d|Y, I) &= \int p(d, \lambda|Y, I) d\lambda \\ &= \int p(d|\lambda, Y, I) p(\lambda|Y, I) d\lambda. \end{aligned} \tag{4.41}$$

To aid our understanding, we make some simplifying assumptions on the two factors in the integral. Suppose we have no prior information on λ and that our model does not favour any value of λ. In this case, the prior on λ is constant within the range $\lambda \in [\lambda_{\text{min}}, \lambda_{\text{max}}]$, i.e.

$$p(\lambda|Y, I) = \frac{1}{\lambda_{\text{max}} - \lambda_{\text{min}}}, \quad \text{for } \lambda_{\text{min}} \leq \lambda \leq \lambda_{\text{max}}. \tag{4.42}$$

Furthermore, we assume the likelihood to be a Gaussian centred around the maximum lilkelihood λ_0 with a standard deviation given by σ_λ. Therefore, the likelihood $p(d|\lambda, Y, I)$ is of the form

$$p(d|\lambda, Y, I) = p(d|\lambda_0, Y, I) \exp\left[-\frac{(\lambda - \lambda_0)^2}{2\sigma_\lambda^2}\right]. \tag{4.43}$$

The evidence is consequently given by

$$\begin{aligned} p(d|Y, I) &= \int p(d|\lambda, Y, I) p(\lambda|Y, I) d\lambda \\ &= \int \frac{1}{\lambda_{\text{max}} - \lambda_{\text{min}}} p(d|\lambda_0, Y, I) \exp\left[-\frac{(\lambda - \lambda_0)^2}{2\sigma_\lambda^2}\right] d\lambda \\ &= \frac{p(d|\lambda_0, Y, I)}{\lambda_{\text{max}} - \lambda_{\text{min}}} \int \exp\left[-\frac{(\lambda - \lambda_0)^2}{2\sigma_\lambda^2}\right] d\lambda \\ &= p(d|\lambda_0, Y, I) \frac{\sigma_\lambda \sqrt{2\pi}}{\lambda_{\text{max}} - \lambda_{\text{min}}}. \end{aligned} \tag{4.44}$$

Finally, the odds ratio can be written as

$$O_Y^X = \frac{p(X|I)}{p(Y|I)} \frac{p(d|X, I)}{p(d|\lambda_0, Y, I)} \frac{\lambda_{\text{max}} - \lambda_{\text{min}}}{\sigma_\lambda \sqrt{2\pi}}. \tag{4.45}$$

The factor $p(X|I)/p(Y|I)$ represents the prior odds of the models. We assume that there is nothing in the background information that favours either X or Y, so that the prior odds equals unity. The factor $p(d|X, I)/p(d|\lambda_0, Y, I)$ compares the best fits of the two models. In the case that Y is a more complicated model compared to X, this factor is usually smaller than unity. If the goodness-of-fit was the only contribution, we would indeed get that the more complicated model is preferred.

However, there is a third contributing factor, given by $(\lambda_{\max} - \lambda_{\min})/(\sigma_\lambda\sqrt{2\pi})$. This factor is big if the prior range for λ is large compared to the width of the likelihood, and conversely small if the width of the likelihood is much smaller than the prior range. If the extra parameter λ is extraneously added such that the width of the likelihood σ_λ is much smaller than the prior range $\lambda_{\max} - \lambda_{\min}$, the third factor penalises model Y. Therefore, one is not at liberty to add arbitrary complexity to the model without being penalised within the formalism of hypothesis testing. This aspect of hypothesis testing is in line with *Occam's razor*, i.e. "it is vain to do with more what can be done with less".

The concepts of hypothesis testing are pivotal when we wish to develop a test for a specific physical theory. For example, it should be clear by now that simply measuring a parameter or comparing best fits cannot constitute a test of the validity of a theory. Furthermore, it should be clear that a test of GR should always compare two generative models. In part II, we show that a generic test of GR can be constructed by considering the GW phase coefficients from a CBC signal, given by Eq. (2.89), and associate propositions to the correctness of these coefficients.

References

1. E. Jaynes, *Probability Theory: The Logic of Science*, Ed. by G. Bretthorst (Cambridge University Press, Cambridge, 2003)
2. D. MacKay, *Information Theory, Inference, and Learning Algorithms*, (Cambridge University Press, Cambridge, 2003)

Chapter 5
Computational Methods

Both parameter estimation and hypothesis testing follow the simple relations Eqs. (4.21) and (4.31), respectively. However, the computation of the posterior and the odds ratio can turn out to be difficult in practice. Suppose we have a problem with N parameters in the model and we want to sample each parameter with an accuracy of 1 part in R. One would need to R^N computations in order to evaluate the posterior and even more to perform the evidence integral. For problems with limited number of parameters, these calculations can be done through brute-force uniform sampling. For problems in higher dimensions, brute-force methods are simply too inefficient to be practically useful. Instead, a range of specialised algorithms are available to alleviate the problems of computation.

In Sect. 5.1, we will introduce the Levenberg-Marquardt algorithm. The Levenberg-Marquardt algorithm aims to find the stationary points of a given function. In Chap. 12, we will use the Levenberg-Marquardt algorithm to find maxima of posterior density functions associated to the cosmological parameters that describe the large scale properties of the Universe.

Furthermore, we will introduce the Fisher information matrix in Sect. 5.2. The Fisher information matrix is used to estimate the accuracy with which parameters in a given model can be measured. The technique is often used in GW data-analysis to obtain a quick estimate of the error associated to a given parameter. In Chap. 12, we will use the Fisher information matrix to estimate the accuracy of the distance measurement by the Einstein Telescope. The accuracy of the distance measurement will then be used to construct likelihood functions for the cosmological parameters.

Finally, in Sects. 5.3 and 5.4 we will look into methods that can approximate the full posterior density function given by Eq. (4.21) and the associated evidence given by Eq. (4.37). Both Markov chain Monte Carlo methods (see Sect. 5.3) and Nested Sampling (see Sect. 5.4) can be used to approximate the posterior density function and the evidence. In Chap. 8, we use these techniques to construct a data-analysis pipeline that aims to test GR through the calculation of the odds ratio given by Eq. (4.32).

T.G.F. Li, *Extracting Physics from Gravitational Waves*,
Springer Theses, DOI 10.1007/978-3-319-19273-4_5

5.1 Levenberg-Marquardt Algorithm

Suppose we want to find the maximum of the posterior, because we are interested in the most likely values of the parameters. One way to tackle this problem is to evaluate the posterior on a grid and find the entry with the highest probability. This method proves to be prohibitive if the dimensionality is more than a few. Instead, we can use our knowledge of the posterior to find the maximum more efficiently.

The stationary points of a function $f(x)$ can be found through the condition $df/dx|_{x0} = 0$. Similarly, if the posterior depends on a set of N parameters $\boldsymbol{\theta}$, we can find the stationary points through the simultaneous equations

$$\partial_{\theta_i} p(\boldsymbol{\theta}|d, H, I)\big|_{\boldsymbol{\theta}_0} = 0, \tag{5.1}$$

where H denotes the generative model and d represents the data. The stationary point is a maximum if the eigenvalues of the Hessian $\partial_{\theta_i}\partial_{\theta_j} p(\boldsymbol{\theta}|d, H, I)\big|_{\boldsymbol{\theta}_0}$ are all negative. Furthermore, if we are largely ignorant about the parameters before the experiment, we can assign a flat prior distribution to the prior $p(\boldsymbol{\theta}|I)$ and write

$$p(\boldsymbol{\theta}|d, H, I) \propto p(d|\boldsymbol{\theta}, H, I). \tag{5.2}$$

Therefore, the maximum posterior coincides with the maximum lilkelihood.

Suppose we perform M independent measurements, yielding d_k for $k = 1, 2, \ldots, M$. The combined likelihood is subsequently given by

$$p(\{d_k\}|\boldsymbol{\theta}, H, I) = \prod_{k=1}^{M} p(d_k|\boldsymbol{\theta}, H, I). \tag{5.3}$$

Similar to Sect. 3.2.1, we assume that the experimental noise is governed by a Gaussian stochastic process with zero mean. This means that the probability distribution of the noise n_k is given by

$$p(n_k|I) = \frac{1}{\sigma_k\sqrt{2\pi}} \exp\left\{-\frac{n_k^2}{2\sigma_k^2}\right\}, \tag{5.4}$$

where σ_k is the standard deviation associated to the characteristics of the experimental noise. We can turn this into a probability distribution for d_k by assuming that $d_k = n_k + f_k$, where the function f relates the parameters $\boldsymbol{\theta}$ to the expectation value of the data, i.e. $f_k(\boldsymbol{\theta}) = f(\boldsymbol{\theta}, k)$. The expression in Eq. (5.4) can then be rewritten as

$$p(d_k|\boldsymbol{\theta}, H, I) = \frac{1}{\sigma_k\sqrt{2\pi}} \exp\left\{-\frac{[d_k - f_k(\boldsymbol{\theta})]^2}{2\sigma_k^2}\right\}, \tag{5.5}$$

Finally, the combined likelihood can be written as

$$p(\{d_k\}|\boldsymbol{\theta}, H, I) \propto \exp\left[-\frac{\chi^2}{2}\right], \tag{5.6}$$

where we have defined the variable χ^2 as

$$\chi^2 \equiv \sum_{k=1}^{N}\left[\frac{d_k - f_k(\boldsymbol{\theta})}{\sigma_k}\right]^2. \tag{5.7}$$

Therefore, the minimisation of χ^2 is equivalent to the maximisation of the likelihood.

5.1.1 Gauss' Method

One way to minimise χ^2 is through Gauss' method. In this method, one expands the generative function f to linear order around $\boldsymbol{\theta}$, i.e.

$$f(\boldsymbol{\theta} + \boldsymbol{\delta\theta}, k) \approx f(\boldsymbol{\theta}, k) + \frac{\partial f_k}{\partial \theta_j}\delta\theta_j, \tag{5.8}$$

where we assume that repeated indices are summed over. Then, Eq. (5.7) can be rewritten as

$$\chi^2 = \left[\frac{d_k - f_k - J_{kj}\delta\theta_j}{\sigma_k}\right]^2, \tag{5.9}$$

where J_{ij} is called the *Jacobian* and is given by

$$J_{ij} = \frac{\partial f_i}{\partial \theta_j}. \tag{5.10}$$

We can simplify the notation by dividing all variables by σ_k and denoting them with a hat, i.e. $\hat{d}_k \equiv d_k/\sigma_k$. To find the minimum, we then solve for the set of simultaneous equations

$$\begin{aligned}\partial_{\delta\theta_i}\chi^2 &= \hat{J}_{ki}\left[\hat{d}_k - \hat{f}_k - \hat{J}_{kj}\delta\theta_j\right] \\ &= 0.\end{aligned} \tag{5.11}$$

Rearranging this equation, we obtain the so-called *normal equation*

$$\hat{J}_{ki}\hat{J}_{kj}\delta\theta_j = \hat{J}_{ki}\left(\hat{d}_k - \hat{f}_k\right). \tag{5.12}$$

Starting from an initial guess $\boldsymbol{\theta}_1$, Eq. (5.12) can be solved for $\boldsymbol{\delta\theta}_1$ and the estimate updated through $\boldsymbol{\theta}_2 = \boldsymbol{\theta}_1 + \boldsymbol{\delta\theta}_1$. The algorithm will thus find the minimum through successive iterations until the Jacobian yields the desired minimisation condition.

Gauss' method has a rapid convergence near the optimal solution. However, solving the normal equation does not guarantee that the updated estimate resides in a location where the approximation Eq. (5.8) is valid. It is therefore not uncommon that Gauss' method fails to converge.

5.1.2 Gradient Method

Another method to find the minimum of a function f is to traverse in the negative direction of the gradient ∇f. Thus, in order to get reach the minimum, we obtain the update on the current state through the gradient of the χ^2 in Eq. (5.7):

$$\begin{aligned}\delta\theta_i &= -\kappa\left(\nabla_{\boldsymbol{\theta}}\chi^2\right)_i \\ &= \kappa\hat{J}_{ki}\left(\hat{d}_k - \hat{f}_k\right),\end{aligned} \tag{5.13}$$

where $\kappa \geq 0$ is a parameter to be adapted throughout the algorithm. Just as Gauss' method, the gradient method finds minima through iteratively updating the initial guess with $\boldsymbol{\delta\theta}$ from Eq. (5.13) until the conditions of a minimum are met.

5.1.3 Levenberg-Marquardt Method

The Levenberg-Marquardt method combines Eqs. (5.12) and (5.13) into the so-called *augmented normal equation*,

$$\left(\hat{J}_{ki}\hat{J}_{kj} - \lambda I_{ij}\right)\delta\theta_j = \hat{J}_{ki}\left(\hat{d}_k - \hat{f}_k\right), \tag{5.14}$$

where I_{ij} is the unit matrix and $\lambda \geq 0$ is again a parameter to be tuned throughout the course of the algorithm. Depending on the value of λ, the algorithm behaves either like Gauss' method or the gradient method. For large λ, the augmented normal equation becomes similar to Eq. (5.13) and thus the gradient method. Conversely, for small λ, the augmented normal equation becomes similar to Eq. (5.12) and thus Gauss' method.

The algorithm decides on the value of λ according to several principles. Firstly, the value of χ^2 should decrease, i.e.

$$\chi^2\left(\boldsymbol{\theta}_{i+1}\right) < \chi^2\left(\boldsymbol{\theta}_i\right). \tag{5.15}$$

Secondly, the value of λ should be kept small enough to make sure that the linear expansion in Eq. (5.8) is sufficient to validate the normal equation in Eq. (5.12).

5.2 Fisher Information Matrix

Suppose we have an experiment with a generative model H, the corresponding set of N parameters $\boldsymbol{\theta}$, the observed data d and the background information I. We can then express the posterior of $\boldsymbol{\theta}$ by using Bayes' theorem

$$p(\boldsymbol{\theta}|d, H, I) = \frac{p(d|\boldsymbol{\theta}, H, I)p(\boldsymbol{\theta}|H, I)}{p(d|H, I)}. \tag{5.16}$$

The only factor on the RHS that depends on the data d and involves the parameters $\boldsymbol{\theta}$ is the likelihood $p(d|\boldsymbol{\theta}, H, I)$. It is therefore evident that

> within the context of the specified model, the likelihood function $p(d|\boldsymbol{\theta}, H, I)$ contains all the information about the set of parameters $\boldsymbol{\theta}$ that is present in the data d.

This statement is often referred to as the *likelihood principle*. Using the likelihood principle, we can argue that in order to assess the ability of an experiment to measure $\boldsymbol{\theta}$, we only need to focus on the likelihood.

5.2.1 Resolving Power of an Experiment

Suppose that the data d consist of a set of M independent measurements

$$d = \{x_1, x_2, \ldots, x_M\}. \tag{5.17}$$

The logarithm of the likelihood function is then given by

$$\frac{1}{M} \ln p(d|\boldsymbol{\theta}, H, I) = \frac{1}{M} \sum_{i=1}^{M} \ln p(x_i|\boldsymbol{\theta}, H, I). \tag{5.18}$$

In the limit of $M \to \infty$, we assume that the relative frequency of the values of x_i is the same as if the values were drawn from the sampling distribution $p(x|\boldsymbol{\theta}_T, H, I)$, where $\boldsymbol{\theta}_T$ is the true value of $\boldsymbol{\theta}$. Therefore, we can assert that

$$\begin{aligned} \lim_{M\to\infty} \frac{1}{M} \ln p(d|\boldsymbol{\theta}, H, I) &= \int p(x|\boldsymbol{\theta}_T, H, I) \ln p(x|\boldsymbol{\theta}, H, I)dx \\ &= E\left[\ln p(x|\boldsymbol{\theta}, H, I)\right]_x . \end{aligned} \tag{5.19}$$

We make the assumption that the true value of $\boldsymbol{\theta}_T$ coincides with the maximum lilkelihood value $\boldsymbol{\theta}_0$. Next, we can expand the likelihood function around the maximum value $\boldsymbol{\theta}_0$

$$\ln p(x|\boldsymbol{\theta}, H, I) \approx \ln p(x|\boldsymbol{\theta}_0, H, I) + \frac{1}{2}\sum_{i,j=1}^{N} \left.\frac{\partial^2 \ln p(x|\boldsymbol{\theta}, H, I)}{\partial\theta_i \partial\theta_j}\right|_{\boldsymbol{\theta}_0} \delta\theta_i\delta\theta_j, \tag{5.20}$$

where we have used the fact that the linear terms in this expansion disappear at the maximum lilkelihood. Taking expectation values, we obtain the following expression

$$E\left[\ln p(x|\boldsymbol{\theta}, H, I)\right]_x \approx E\left[\ln p(x|\boldsymbol{\theta}_0, H, I)\right]_x + \frac{1}{2}\sum_{i,j=1}^{N} E\left[\left.\frac{\partial^2 \ln p(x|\boldsymbol{\theta}, H, I)}{\partial\theta_i \partial\theta_j}\right|_{\boldsymbol{\theta}_0}\right]_x \delta\theta_i\delta\theta_j. \tag{5.21}$$

Rearranging this expression and using Eq. (5.19) we can write

$$\lim_{M\to\infty} \frac{1}{M} \ln\left[\frac{p(d|\boldsymbol{\theta}, H, I)}{p(d|\boldsymbol{\theta}_0, H, I)}\right] = -\frac{1}{2}\sum_{i,j=1}^{N} \mathcal{I}_{ij}\delta\theta_i\delta\theta_j, \tag{5.22}$$

where the *Fisher information matrix* $\mathcal{I}_{ij}$ is defined as

$$\begin{aligned} \mathcal{I}_{ij} &\equiv -E\left[\left.\frac{\partial^2 \ln p(x|\boldsymbol{\theta}, H, I)}{\partial\theta_i \partial\theta_j}\right|_{\boldsymbol{\theta}_0}\right]_x \\ &= -\int p(x|\boldsymbol{\theta}_0, H, I) \left.\frac{\partial^2 \ln p(x|\boldsymbol{\theta}, H, I)}{\partial\theta_i \partial\theta_j}\right|_{\boldsymbol{\theta}_0} dx. \end{aligned} \tag{5.23}$$

Equation (5.22) already gives us a hint that the Fisher information matrix is a measure of an experiment's resolving power for the parameters $\boldsymbol{\theta}$. Formally, the Fisher information matrix can be shown to follow the *Cramer-Rao bound* [1, 2]

$$\sigma^2(\theta_i) \geq \left(\mathcal{I}^{-1}\right)_{ii}, \tag{5.24}$$

i.e. the diagonal elements of the inverse Fisher matrix give a lower bound on the variances expected from an experiment.

5.2.2 Fisher Matrix in Gravitational-Wave Data Analysis

Suppose that the detector output is given by

$$s(t) = h(t, \boldsymbol{\theta}) + n(t), \tag{5.25}$$

where $n(t)$ is the noise described by a Gaussian stochastic process, *cf.* Sect. 3.2.1, and $h(t, \boldsymbol{\theta})$ denotes the GW which depends on the set of N parameters $\boldsymbol{\theta}$. The probability distribution of the detector output is given by

$$\begin{aligned} p(s|\boldsymbol{\theta}) &\propto \exp\left\{-\frac{1}{2}\int dt_1 dt_2\; n(t_1)\Omega(t_1, t_2)n(t_2)\right\} \\ &\propto \exp\left\{-\frac{1}{2}\int dt_1 dt_2\; [s(t_1) - h(t_1; \boldsymbol{\theta})]\,\Omega(t_1, t_2)\,[s(t_2) - h(t_2; \boldsymbol{\theta})]\right\}, \end{aligned} \tag{5.26}$$

where we have replaced the auto-correlation function R from Eq. (3.10) with

$$\Omega(t_1, t_2) \equiv R^{-1}(t_1, t_2). \tag{5.27}$$

Consequently, the Fisher matrix is given by

$$\begin{aligned} \mathcal{I}_{ij} &\equiv -E\left[\left.\frac{\partial^2 \ln p(s|\boldsymbol{\theta}, H, I)}{\partial\theta_i \partial\theta_j}\right|_{\boldsymbol{\theta}_0}\right]_s \\ &= -\int \mathcal{D}s dt_1 dt_2\; p(s|\boldsymbol{\theta}_0, H, I)\left\{\frac{\partial^2 h}{\partial\theta_i\partial\theta_j}(t_1; \boldsymbol{\theta}_0)\Omega(t_1, t_2)\,[s(t_2) - h(t_2; \boldsymbol{\theta}_0)]\right. \\ &\left. -\frac{\partial h}{\partial\theta_i}(t_1; \boldsymbol{\theta}_0)\Omega(t_1, t_2)\frac{\partial h}{\partial\theta_j}(t_2; \boldsymbol{\theta}_0)\right\}, \end{aligned} \tag{5.28}$$

where $\int \ldots \mathcal{D}s$ denotes the functional integral over all possible data realisations. Next, we assume that the SNR is high and that first term on the RHS can be neglected compared to the second term. We can thus write

$$\begin{aligned} \mathcal{I}_{ij} &\approx \int dt_1 dt_2 \left\{\frac{\partial h}{\partial\theta_i}(t_1; \boldsymbol{\theta}_0)\Omega(t_1, t_2)\frac{\partial h}{\partial\theta_j}(t_2; \boldsymbol{\theta}_0)\right\}\int \mathcal{D}s\; p(s|\boldsymbol{\theta}_0, H, I) \\ &\approx \int dt_1 dt_2 \left\{\frac{\partial h}{\partial\theta_i}(t_1; \boldsymbol{\theta}_0)\Omega(t_1, t_2)\frac{\partial h}{\partial\theta_j}(t_2; \boldsymbol{\theta}_0)\right\} \\ &\approx \int df_1 df_2 \left\{\frac{\partial \tilde{h}}{\partial\theta_i}(f_1; \boldsymbol{\theta}_0)\tilde{\Omega}(-f_1, -f_2)\frac{\partial \tilde{h}}{\partial\theta_j}(f_2; \boldsymbol{\theta}_0)\right\}. \end{aligned} \tag{5.29}$$

From Eqs. (3.11) and (5.27), we can verify that

$$\tilde{\Omega}(f_1, f_2) = \frac{\delta(f_1 + f_2)}{\frac{1}{2}S_n(f_1)}. \tag{5.30}$$

Inserting Eq. (5.30) into Eq. (5.29), we have

$$\begin{aligned}
\mathcal{I}_{ij} &\approx \int df_1 df_2 \left\{ \frac{\partial \tilde{h}}{\partial \theta_i}(f_1;\boldsymbol{\theta_0}) \frac{\delta(-f_1-f_2)}{\frac{1}{2}S_n(-f_1)} \frac{\partial \tilde{h}}{\partial \theta_j}(f_2;\boldsymbol{\theta_0}) \right\} \\
&\approx \int df_2 \left\{ \frac{\partial \tilde{h}}{\partial \theta_i}(-f_2;\boldsymbol{\theta_0}) \frac{1}{\frac{1}{2}S_n(f_2)} \frac{\partial \tilde{h}}{\partial \theta_j}(f_2;\boldsymbol{\theta_0}) \right\} \\
&\approx \int df_2 \left\{ \frac{\partial \tilde{h}}{\partial \theta_i}^{*}(f_2;\boldsymbol{\theta_0}) \frac{1}{\frac{1}{2}S_n(f_2)} \frac{\partial \tilde{h}}{\partial \theta_j}(f_2;\boldsymbol{\theta_0}) \right\},
\end{aligned} \tag{5.31}$$

where we have used the fact that $h(t)$ is real. Finally, with the definition of the inner product in Eq. (3.22), we can write

$$\mathcal{I}_{ij} = \left(\frac{\partial h}{\partial \theta_i} \middle| \frac{\partial h}{\partial \theta_j} \right) \bigg|_{\boldsymbol{\theta_0}} . \tag{5.32}$$

Therefore, the ability for GW detectors to measure parameters can be estimated by the knowledge of the waveform and the power spectral density.

5.3 Monte Carlo Methods

What we have seen so far are methods to obtain specific pieces of information from a target distribution, e.g. Levenberg-Marquardt algorithm to find the maximum and the Fisher information matrix for the variance. To get a complete picture, one would like to obtain the full probability (density) function, $P(x)$. However, it is often the case that one only has access to a function $\tilde{P}(x)$, where

$$P(x) = \frac{\tilde{P}(x)}{Z}, \tag{5.33}$$

i.e. one does not have access to the normalisation constant Z. Furthermore, even if one has access to Z, evaluating $P(x)$ in a high-dimensional space can be cumbersome, even for modern computers.

To alleviate this problem, one can resort to the use of Monte Carlo methods. Monte Carlo algorithms are computational techniques that use random numbers to generate samples from a desired probability distribution $P(x)$. The goal is thus to generate a set of samples $\{x^{(r)}\}$ from $\tilde{P}(x)$.

5.3.1 Markov Chain Monte Carlo Methods

MCMC methods form a subclass of Monte Carlo methods that sample the desired distribution $P(x)$ by using an object that transitions from one location of the parameter space to another. An MCMC method is characterised by an initial distribution $p^{(0)}(x)$ and a transition probability $T(x', x)$. The probability distribution of the state at the $t + 1$th iteration of the Markov chain is given by

$$p^{(t+1)}(x') = \int dx\ T(x', x) p^{(t)}(x). \tag{5.34}$$

The Markov chain must posses two properties [3]. Firstly the desired distribution $P(x)$ is an *invariant* distribution of the chain. The invariance of a distribution $\pi(x)$ under the transformation $T(x', x)$ is given by

$$\pi(x') = \int dx\ T(x', x)\pi(x). \tag{5.35}$$

Alternatively, the chain must satisfy *detailed balance*, i.e.

$$T(x_a, x_b)P(x_b) = T(x_b, x_a)P(x_a). \tag{5.36}$$

Secondly, the chain must be ergodic, i.e. tend to an invariant distribution $\Lambda(x)$ under the transformation $T(x', x)$. This can be written as

$$p^{(t)}(x) \to \Lambda(x) \text{ as } t \to \infty \text{ for any } p^{(0)}(x). \tag{5.37}$$

5.3.2 Metropolis-Hasting Sampling

The Metropolis-Hastings algorithm is an MCMC method that aims to sample from a target distribution that is inherently difficult to sample by using a different distribution from which we do know how to sample.

The Metropolis-Hastings algorithm works as follows. A new sample $x^{(t+1)}$ is generated from the proposal distribution $Q(x; x^{(t)})$. The transition from $x^{(t)}$ to $x^{(t+1)}$ depends on the *acceptance* a, where

$$a = \frac{P(x')}{P(x^{(t)})} \frac{Q(x^{(t)}; x')}{Q(x'; x^{(t)})}. \tag{5.38}$$

If the proposal distribution is symmetric, such as for a Gaussian distribution, the acceptance simply becomes

$$a = \frac{P(x')}{P(x^{(t)})}. \tag{5.39}$$

Subsequently, the new sample $x^{(t+1)}$ is chosen according to

$$\text{if a} \geq 1: \quad x^{(t+1)} = x'$$
$$\text{else:} \quad x^{(t+1)} = \begin{cases} x' & \text{with probability } a \\ x^{(t)} & \text{with probability } 1 - a. \end{cases} \tag{5.40}$$

The disadvantage of the Metropolis-Hastings algorithm is that it produces correlated samples. Hence, samples should only be kept for every n iterations, such that n steps in the proposal distribution are larger than the auto-correlation length of the proposal distribution. As an advantage, the proposal distribution does not need to match the target distribution in order for the Metropolis-Hastings algorithm to be practically useful. Therefore, the Metropolis-Hastings algorithm is a generically useful approach to sample distributions such as the posterior or the prior.

5.4 Nested Sampling

We now turn to the calculation of the evidence

$$\begin{aligned} p(d|H,I) &= \int d^N\boldsymbol{\theta}\; p(d|\boldsymbol{\theta},H,I)p(\boldsymbol{\theta}|H,I) \\ &= \int d^N\boldsymbol{\theta}\; L(\boldsymbol{\theta})\pi(\boldsymbol{\theta}), \end{aligned} \tag{5.41}$$

where $L(\boldsymbol{\theta})$ represents the likelihood and $\pi(\boldsymbol{\theta})$ represents the prior. The calculation of the evidence is crucial in performing hypothesis testing, *cf.* Sect. (4.3). In this section, we describe *Nested Sampling*, which is an algorithm aimed at efficiently computing the evidence integral Eq. (5.41), even if the parameter space is large or has a high dimensionality. As an optional by-product, Nested Sampling also outputs the posterior distribution at minimal extra computational cost.

5.4.1 *Basic Idea*

Nested Sampling computes the evidence by rewriting the functions in Eq. (5.41) that depend on the set of N parameters $\boldsymbol{\theta}$ in terms of a single scalar variable called the *prior mass* X. This prior mass is defined as "the fraction of the prior volume with likelihood greater than λ". Mathematically, the prior mass is defined as

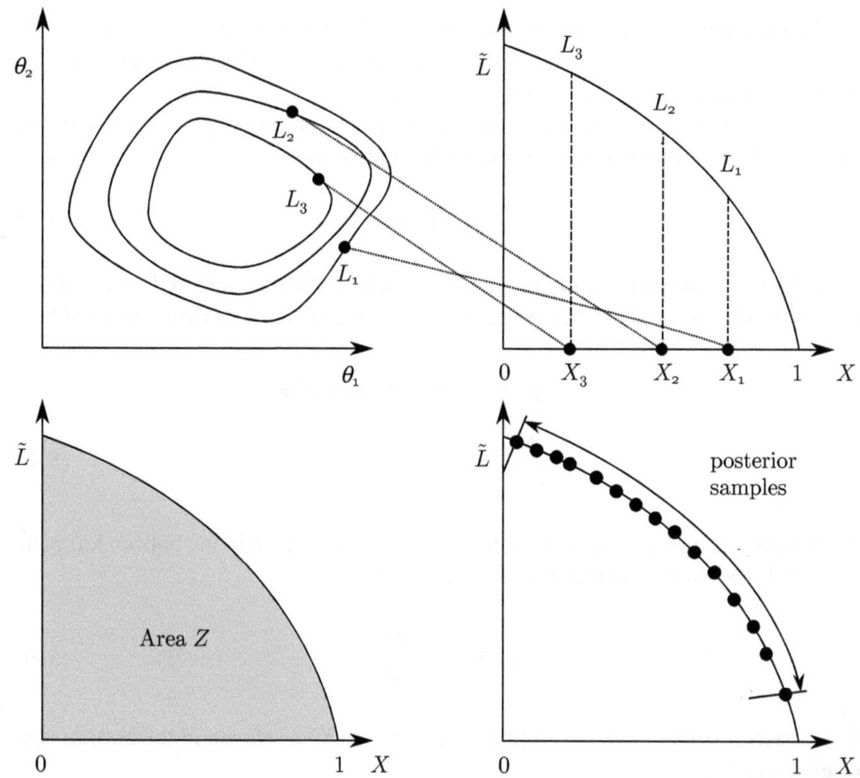

Fig. 5.1 Schematic example of the transformation of the evidence integral from a multi-dimensional integral into a single dimensional one. The *top left panel* shows the equal likelihood contours in the (θ_1, θ_2) plane of the parameter space Θ. The *black lines* represent the contours for equal-likelihood and the area inside represents the prior probability subject to the likelihood constraint $L > L_i$, the prior mass. The *top right panel* shows the likelihood as a function of the prior mass $\tilde{L}(X)$. The *bottom left panel* shows that the evidence is simply given by the area under the $\tilde{L}(X)$ curve. The *bottom right panel* is a depiction of posterior samples obtained by sampling the $\tilde{L}(X)$ curve

$$X(\lambda) \equiv \int \int \cdots \int_{L(\boldsymbol{\theta})>\lambda} \pi(\boldsymbol{\theta}) d^N \boldsymbol{\theta}. \tag{5.42}$$

In other words, the prior mass corresponds to the prior integrated over the hypervolume in the N-dimensional parameter space Θ which is bounded by the hypersurface characterised by $L = \lambda$, as show in the top left panel of Fig. 5.1. The element of the prior mass is given by

$$dX = \pi(\boldsymbol{\theta}) d^N \boldsymbol{\theta}. \tag{5.43}$$

Since the prior probability density function $\pi(\boldsymbol{\theta})$ is normalised with respect to $\boldsymbol{\theta}$, the prior mass has a range of $X \in [0, 1]$. The lower bound $X = 0$ corresponds to the surface within which there is no higher likelihood, thus associated to $\lambda = L_{\max}$. The upper bound $X = 1$ corresponds to the surface where all the points inside have a higher likelihood, and is associated to $\lambda = L_{\min}$.

We can now transform the factors in Eq. (5.41) to functions depending on the prior mass. Firstly, the likelihood function can be expressed as

$$\tilde{L}(X(\lambda)) \equiv \lambda. \tag{5.44}$$

where $\tilde{L}(X)$ is a monotonically decreasing function. An example of $\tilde{L}(X)$ is shown in the top right panel of Fig. 5.1. Secondly, we can rewrite the evidence integral to be

$$\begin{aligned} Z &= \int\int\cdots\int L(\boldsymbol{\theta})\pi(\boldsymbol{\theta})d^N\boldsymbol{\theta} \\ &= \int \tilde{L}(X)dX. \end{aligned} \tag{5.45}$$

The evidence is thus the area under the $\tilde{L}(X)$ curve as shown in the bottom left panel of Fig. 5.1. Lastly, the posterior $P(\boldsymbol{\theta})$ is given by

$$\tilde{P}(X) = \frac{\tilde{L}(X)}{Z}. \tag{5.46}$$

This posterior is obtained by sampling the curve $\tilde{L}(X)$ as shown in the bottom right panel of Fig. 5.1.

The idea behind Nested Sampling is thus to construct the function $\tilde{L}(X)$. This is done by progressively finding locations in the parameter space with higher likelihood and assigning a progressively smaller prior mass to these locations (as $\tilde{L}(X)$ is a monotonically decreasing function). Once $\tilde{L}(X)$ has been mapped, the assignment of the evidence and the posterior can be done with Eqs. (5.45) and (5.46). Before we explain the Nested Sampling algorithm further, we can elucidate the concepts thus far with two examples from Skilling [4].

Example 1

Suppose we have a two dimensional parameter space and the corresponding likelihood function, represented by a 4×4 grid

$$L = \begin{bmatrix} 0 & 8 & 15 & 3 \\ 11 & 24 & 22 & 10 \\ 19 & 30 & 26 & 16 \\ 9 & 23 & 18 & 6 \end{bmatrix}. \tag{5.47}$$

To each cell of the 16 cells, we assign an equal prior mass element of $\Delta X = 1/16$. The idea of Nested Sampling is to sort the likelihood elements starting with the lowest value of the likelihood, which in this case is given by

$$L = (0, 3, 6, 8, 9, 10, 11, 15, 16, 18, 19, 22, 23, 24, 26, 30). \tag{5.48}$$

To get the evidence, one approximates the integral in Eq. (5.45) by

$$\begin{aligned} Z &= \int \tilde{L}(X) dX \\ &\approx \sum_i \tilde{L}_i \Delta X_i, \end{aligned} \tag{5.49}$$

and evaluates the evidence to be

$$Z = \frac{0+3+6+8+9+10+11+15+16+18+19+22+23+24+26+30}{16} = 5. \tag{5.50}$$

Having sorted the two dimensional likelihood into a one dimensional array, we can now approximate the likelihood for $X = 1/5$ by the fourth item (roughly a fifth of the sixteen items) counting from the right, from the sorted list. The associated likelihood is thus $\tilde{L}(X = 1/5) \approx 23$ (remember that $X = 1/5$ confines a fifth of the total prior volume in which the likelihood is greater than the likelihood of the boundary). The corresponding area in parameter space is shown in grey

$$L = \begin{bmatrix} 0 & 8 & 15 & 3 \\ 11 & 24 & 22 & 10 \\ 19 & 30 & 26 & 16 \\ 9 & 23 & 18 & 6 \end{bmatrix}. \tag{5.51}$$

Example 2

In the second example, we do not assume that the likelihood has already been mapped. Suppose we first drop 3 so-called *live points* uniform with respect to the prior. On the left panel of Fig. 5.2 these points are denoted by the numbers 1, 3, 4. Their associated prior mass X_i is shown in the right panel of Fig. 5.2. Point 1 has the lowest likelihood L_1 and the highest prior mass X_1. We assume for the moment that we know how to assign the prior mass according to a given likelihood. Finally, we store its location in the parameter space $\boldsymbol{\theta}_1$, its prior mass X_1 and its likelihood L_1.

Next, we discard the point with the lowest likelihood (point 1) from our set of live points and sample a new point from the prior, but now with the constraint that $L > L_1$. This new point is denoted by point 2. Note that the remaining live points (points 3 and 4) automatically satisfy the condition $L > L_1$. Again, we find the live point with the smallest likelihood, which turns out to be point 2 (note that it is not necessary that the newly drawn point has the lowest likelihood value; it only has to obey the condition $L > L_1$). Since $L_2 > L_1$, we know that the prior mass for point 2 must satisfy $X_2 < X_1$. As with X_1, the parameters of point 2 are stored and the point is subsequently discarded from the set of live points in favour of again a new point, but now subject to the adjusted likelihood condition $L > L_2$. After 5 iterations, we have collected 5 sorted points (1, 2, 3, 4, 5) from the $\tilde{L}(X)$ curve, which allows us the perform the integral in Eq. (5.45) and obtain an estimation of the evidence.

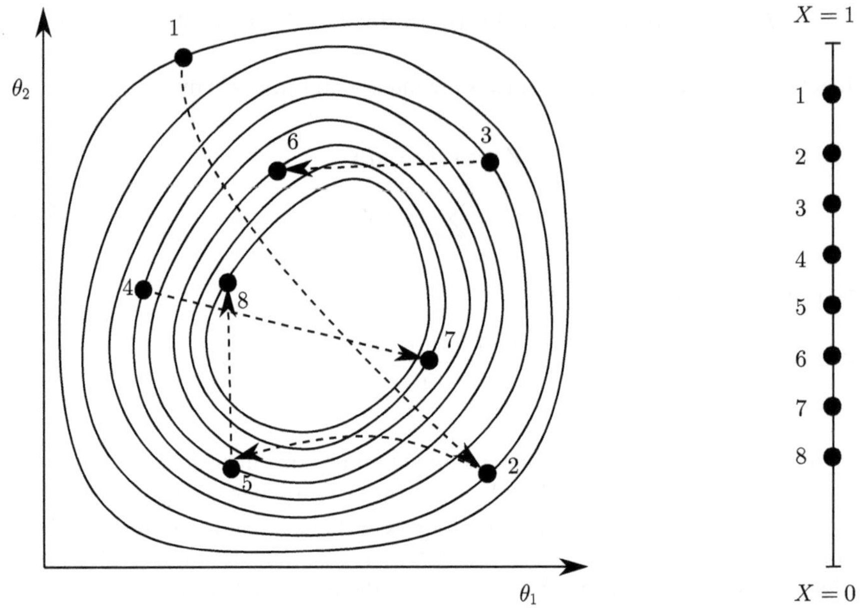

Fig. 5.2 Schematic overview of the Nested Sampling algorithm with 3 so-called *live points* and 5 iterations. The *left panel* shows the locations in the parameter space which either of the three live points have explored. The *right panel* shows these same locations, but now in terms of the prior mass

5.4.2 Algorithm

Probabilistic Estimation of the Prior Mass

Although there is a correspondence between the parameter space Θ and the prior mass X, the mapping might not be trivially done through brute-force computation if the parameter space is large or if the dimensionality is high. Instead, Nested Sampling seeks to assign the prior mass statistically.

Suppose we have two points in the parameter space $\boldsymbol{\theta_0}$, $\boldsymbol{\theta_1}$, their corresponding likelihoods L_0, L_1 and prior masses X_0, X_1, respectively. If $L_1 > L_0$, then, from the definition of the prior mass in Eq. (5.42), we can assert that

$$X_0 > X_1 \text{ for } L_0 < L_1. \tag{5.52}$$

Furthermore, from Eq. (5.43), we know that the density of states of the prior mass is equal to the density of states of the prior in the parameter space. Therefore, if we sample the prior in the parameter space subject to the constraint $L > L^*$, we know we are sampling the prior mass uniformly between $0 < X < X^*$, where X^* is the prior mass corresponding the likelihood value L^*. With this in mind, we can now statistically assign a prior mass corresponding to the exploration of our parameter space.

Suppose we drop M samples across the entire parameter space, labelled by an index i, according to the prior probability distribution. Each of these samples has a likelihood associated with it through its location in the parameter space. Thus, we know that each of these samples is associated to a surface within which the likelihood is higher than at the boundary. Equivalently, we know that these samples are uniformly sampled in the entire prior mass region (between 0 and 1). Therefore, each sample in prior mass can be considered to be drawn from

$$X \in U(0, 1), \tag{5.53}$$

where $U(0, 1)$ represents the uniform distribution in the interval $[0, 1)$. The probability that the prior mass is drawn with a value less than some X is given by

$$\begin{aligned} P(X) &= \int_0^X dX' \\ &= X. \end{aligned} \tag{5.54}$$

If we define the X^* to be the highest prior mass, then we know this surface has a corresponding likelihood L^* that is the smallest of the set of M likelihoods. The probability that the surface with the highest prior mass is at $X = \chi$ is therefore given by the joint probability that none of the samples have a prior mass greater than χ. This joint probability distribution is thus

$$\begin{aligned} P(\{X_i\} < \chi) &= \prod_i^M \int_0^\chi dX_i \\ &= \prod_i^M \chi \\ &= \chi^M. \end{aligned} \tag{5.55}$$

The probability density that the highest of M samples has a prior mass of χ is subsequently given by

$$\begin{aligned} P(\chi, M) &= \frac{\partial P(X_i < \chi)}{\partial \chi} \\ &= M\chi^{M-1}. \end{aligned} \tag{5.56}$$

Similarly, if we sample M points from the prior within the region $0 < X < X^*$, we can define the *shrinkage ratio* $t \equiv \frac{X}{X^*}$. The shrinkage ratio follows the same distribution as χ in Eq. (5.56)

$$P(t, M) = Mt^{M-1}, \tag{5.57}$$

and the prior mass X can then be obtained by $X = tX^*$.

Sampling the Constrained Prior

The main difficulty of Nested Sampling is sampling the prior subject to the constraint $L > L^*$. For most problems, the prior cannot be straightforwardly sampled and one has to resort to Monte Carlo methods.

The specific details of sampling the constrained prior are unimportant to the Nested Sampling algorithm, as long as one is able to obtain (approximately) independent samples. It therefore suffices to use a Monte Carlo method. A common method used in conjunction with Nested Sampling is the Metropolis-Hastings algorithm, explained in Sect. 5.3.

Termination Condition

The steps described above can, in principle, be repeated *ad infinitum* and can be used to acquire an increasingly better estimate of the evidence. In practice, one would like to have a termination condition that ensures a good evidence estimation in the least amount of time.

Although there is no obvious termination condition associated to Nested Sampling, several practical guidelines can be applied. Firstly, one can estimate the *information* as a function of the evidence and the likelihood

$$\begin{aligned} \mathcal{H} &= \int P(X) \ln (P(X))\, dX \\ &\approx \sum_k \frac{L_k}{Z} \ln \frac{L_k}{Z} \Delta X_k, \end{aligned} \tag{5.58}$$

where $P(X)$ is the posterior density as a function of the prior mass. As most of the evidence is found within $X = e^{-\mathcal{H}}$, one can terminate the algorithm when most of the information is acquired [3].

Secondly, one can estimate the amount of evidence yet to be accumulated and compare that to the evidence already accumulated. The algorithm can be stopped if the remaining evidence is a user-specified fraction of the evidence already accumulated. For example, we can assume that all the remaining points have a likelihood equal to the maximum lilkelihood in the set of live points L_{max}. The termination condition then becomes

$$L_{\text{max}} X_{\text{cur}} < \alpha Z_{\text{cur}}, \tag{5.59}$$

where α is a user-specified constant.

Ultimately, the termination condition will depend on the specifics of the likelihood and the prior. Therefore, a termination condition is usually found through trial and error, satisfying the required accuracy as well as optimising for the computation time.

Computing the Posteriors

Although Nested Sampling is mainly aimed at producing the evidence, it can also estimate the posterior as a by-product. Recall that Nested Sampling stores a set of locations $\{\boldsymbol{\theta}_k\}$ in the parameter space, the associated likelihoods $\{L_k\}$ and prior masses $\{X_k\}$. Each of the samples $\boldsymbol{\theta}_i$ obtained from the Nested Sampling algorithm occupies a fraction of the posterior distribution given by the associated fraction of prior volume and the likelihood value at $\boldsymbol{\theta}_i$,

$$w_i = \frac{L(\boldsymbol{\theta}_i)\Delta X_i}{Z}. \tag{5.60}$$

This fraction of the posterior w_i is then the weight given to each sample in order to construct the posterior distribution.

Nested Sampling in Pseudo-code

We can summarise the Nested Sampling algorithm in terms of pseudo-code. We assume the algorithm to run with M live points and that the sampling of the prior with a likelihood constraint does not pose computational problems.

1. Sample M points $\boldsymbol{\theta}_1, \ldots, \boldsymbol{\theta}_M$ from prior $\pi(\boldsymbol{\theta})$.
2. While not termination condition
   (a) record live point i with lowest L_i as L_k,
   (b) assign $X_k = t_k X_{k-1}$ where t_k from $P(t_k) = M t_k^{M-1}$,
   (c) replace point i with sample from $\pi(\boldsymbol{\theta})$ subject to $L_i > L_k$.
3. Estimate Z by integrating $\{L_k, X_k\}$.

5.4.3 Accuracy of Nested Sampling

There are several uncertainties that can be associated to the probabilistic nature of the Nested Sampling algorithm. Firstly, the probabilistic determination of the prior mass introduces an uncertainty that is linked to the spread of the distribution in Eq. (5.57).

Suppose we use a series of shrinkage factors denoted by $\boldsymbol{t}$. One can quantify the statistical properties of the evidence and the posterior by evaluating its jth moment

$$E\left[Z^j\right]_t = \int [Z(\boldsymbol{t})]^j \, P(\boldsymbol{t})\mathcal{D}\boldsymbol{t}, \tag{5.61}$$

$$E\left[(P(\theta_k))^j\right]_t = \int (P(\theta_k, \boldsymbol{t}))^j \, P(\boldsymbol{t})\mathcal{D}\boldsymbol{t}. \tag{5.62}$$

Computationally, we can achieve this by generating the series of shrinkage factors a number of times with a different state of the random number generator. With a collection of series of shrinkage factors, we can obtain a collection of evidences from which we can obtain the statistical properties. Note that this does not require us to rerun the algorithm as the likelihood values are obtained independent of the values of the prior mass.

As it turns out, this contribution to the uncertainty decreases with the number of live points used. This can be seen by evaluating the variance of Eq. (5.57), given by

$$\sigma_t^2 = \frac{M}{(M+1)^2\,(M+2)}. \tag{5.63}$$

We see that the uncertainty goes to zero in the limit of large M. Therefore, providing a sufficient number of live points reduces the uncertainty in the distribution of t.

Secondly, the imperfect sampling of the prior is a source of uncertainty. This kind of uncertainty affects the set of likelihoods from our live points and ultimately affects the assignment of the prior mass. However, this uncertainty depends on the details of the method used to sample the prior. A quantification of this source of uncertainty can be obtained by running the algorithm a number of times with a different state of the random number generator.

The accuracy of Nested Sampling will ultimately depend on the problem at hand. The choice for the number of live points and the method used to sample the prior depends largely on the shape and dimensionality of the likelihood and prior.

5.4.4 Nested Sampling in Gravitational-Wave Data Analysis

Consider a data stream s that represents the strain output of the interferometer. The data stream can consist of noise, n, and the GW signal, h. We can compute the evidence for the following two hypotheses.

1. $\mathcal{H}_N$ is the noise-only model. It assumes that the detector strain only results from noise sources, i.e.

$$s = n. \tag{5.64}$$

   This model has no free parameters, as it is only a function of the noise.
2. $\mathcal{H}_S$ is the signal model. This model assumes that the detector response consists of the true GW signal embedded in noise, i.e.

$$s = n + h_S. \tag{5.65}$$

   This hypothesis does depend on the signal parameters $\boldsymbol{\theta}_S$ through h_S.

To obtain the desired odds ratio, one needs to compute the evidences $P(s|\mathcal{H}_i, \mathrm{I})$ for each of the two aforementioned hypotheses [*cf.* Eq. (4.32)]. Assuming that the noise is governed by a Gaussian stochastic process [see Eq. (5.4)], the evidence for the noise-only hypothesis $\mathcal{H}_N$ is given by

$$P(s|\mathcal{H}_N, \mathrm{I}) \propto e^{-(s|s)/2}, \tag{5.66}$$

For the signal hypothesis $\mathcal{H}_S$, the evidence is given by

$$P(s|\mathcal{H}_S, \mathrm{I}) \propto \int d^N\boldsymbol{\theta}\, e^{-[s-h(\boldsymbol{\theta})|s-h(\boldsymbol{\theta})]/2}\pi(\boldsymbol{\theta}), \tag{5.67}$$

where the model waveform h can be different from the signal waveform h_S. For example, a real GW will be different from a waveform obtained by using the PN formalism. In simulations, such differences can be introduced by using different approximations of the waveform. The evidence integral can be computed with the Nested Sampling algorithm. However, one still needs to choose the appropriate prior probability function for the GW parameters. The choice of priors are left to be discussed when the method parameters are chosen for the simulations (see Sect. 8.1.2).

Finally, the joint likelihood for the coherent analysis of the output from multiple detectors is simply the product of the individual likelihoods

$$P(s_{\mathrm{coh}}|\mathcal{H}_i, \mathrm{I}) = \prod_j P(s_j|\mathcal{H}_i, \mathrm{I}), \tag{5.68}$$

where the index j runs over the individual detectors considered, and the $\mathcal{H}_i$ is either the noise or signal hypothesis.

References

1. H. Cramér, *Methods of Mathematical Statistics* (Princeton University Press, Princeton, 1946), p. 23
2. C. Rao, Information and the accuracy attainable in the estimation of statisticalparameters. Bull. Calcutta Math. Soc. **37**, 81–91 (1945)
3. D. MacKay, *Information Theory, Inference, and Learning Algorithms* (Cambridge University Press, Cambridge, 2003)
4. J. Skilling, Nested sampling for general Bayesian computation. Bayesian Anal. **1**(4), 833–860 (2006)

Part II
Testing the Strong-Field Dynamics of General Relativity

Chapter 6
Introduction

6.1 Classic Tests

Einstein's theory of GR is known for its mathematical elegance, but the theory also acquired great success through its agreement with experiments. Three tests of GR are considered to be the classic tests. The first classic test was the correct explanation of the anomalous perihelion advance of Mercury. Despite careful analyses, Newton's laws could not account for an excess of 43 arc seconds per century in the perihelion shift, discovered as early as 1845 by Le Verrier [1]. In 1915, Einstein showed that his theory of GR could explain this anomalous shift, without the need of an ad hoc solution such as an extra planet [2].

The second classic test was the measurement of the deflection of light by the Sun. As light grazes the surface of the Sun it is deflected by the mass of the Sun. Based on Newtonian arguments, this deflection was calculated to be 0.88 arc seconds, whereas GR predicted this to be 1.75 arc seconds. Two teams led by Eddington measured the positions of stars before and during the solar eclipse of 1919. The measurements of the angular displacement by the two groups confirmed the prediction by GR [3]. However, these results were later met with skepticism over systematic errors. Nevertheless, upon the announcement of the results, Einstein and his theory of GR became world famous.

The third test, the measurement of the gravitational redshift, was already hypothesised by Einstein in 1907. It was not until 1960 that Pound and Rebka measured the redshift associated to the non-uniformity of the gravitational field [4]. In this experiment, photons were sent down the 74 ft tower of the Jefferson Physical Laboratory (Harvard University) and their frequency was measured at the ground. The resulting shift in frequency was consistent, up to a few percent in accuracy, with Einstein's prediction that photons gain energy, i.e. are measured with a higher frequency, as they go down the Earth's gravitational potential. Similarly, photons were found to lose energy when they were sent upwards. This experiment confirmed the influence of a massive object on spacetime itself. The measurement of the gravitational redshift sparked a period of precision tests of GR.

T.G.F. Li, *Extracting Physics from Gravitational Waves*,
Springer Theses, DOI 10.1007/978-3-319-19273-4_6

6.2 Precision Tests

As technology improved over the years, new tests of GR were devised in order to put Einstein's theory under scrutiny. Most of these tests could not be performed in a regular laboratory, as the distances were too short and objects not massive enough. Instead, the solar system and beyond became the test bed for GR.

The sheer volume of different precision tests that were performed makes it impossible to review them all. However, below is a list of a few notable tests.

- Shapiro time delay: As radar signals graze a massive object, they experience a time delay in a round-trip due to the gravitational potential [5]. The most accurate test on this phenomenon was done by the Cassini-Huygens spacecraft and its result is in accordance with the predictions made by GR to less than 2×10^{-3} % [6].
- Gravitational lensing: Similar to the deflection of light by the Sun, gravitational lensing occurs through the deflection of light by massive objects. Such massive objects, e.g. galaxies or galaxy clusters, act like a lens for light, magnifying and distorting the images observed here on Earth [7]. Measurements of the deflection of radio waves by the solar gravitational field using a technique called very-long baseline interferometry (VLBI) constrain GR to $\sim 1 \times 10^{-2}$ %. [8].
- Nordtvedt effect: If the Earth and the Moon were to have different fractional contributions from self-gravitation to the mass, i.e. the equivalence principle is violated, they would each behave differently under the Sun's gravitational field, causing a shift of the orbit between the Earth and the Moon [9]. The most up to date measurement of this effect is done through laser observation on the Earth-Moon system [10]. The so-called Nordtvedt parameter is found to be $\eta = (-0.6 \pm 5.2) \times 10^{-4}$, where $\eta = 0$ corresponds to GR.
- Lense-Thirring effect: GR predicts the existence of a relativistic correction to the precession of a gyroscope near a large rotating mass, called the Lense-Thirring effect [11]. Gravity Probe B reported a geodetic drift rate and a frame-dragging drift rate that are consistent with the predictions by GR up to $\sim 6 \times 10^{-2}$ % and ~ 5 % in accuracy, respectively [12].

6.3 Binary Pulsars

As impressive as the solar system tests of GR have been on the technological side, the regime in which GR was tested does not compare with the fully dynamical and non-linear aspects that are predicted by GR. The measurements of GWs could be the primary means through which experiments will have access to these aspects of GR.

A first realisation of such measurements, albeit indirect, came from the discovery of the binary pulsar system PSR B1913 + 16 by Hulse and Taylor [13]. This binary system consists of a pulsar and a companion NS, spiralling around each other. As the two objects go through their orbital motion around their common centre of mass, the orbital separation decreases as GWs carry away energy and angular momentum.

Through precise measurements on the pulsar, one can infer the system's characteristics, and in particular, the change in the orbital period. As it turns out, this change is in close agreement with the quadrupole emission of GWs, described in Sect. 1.6.2, and leaves little room to doubt the quadrupole formula for the emission of GWs.

In order to gauge the regime in which tests of GR reside, it is instructive to look at two dimensionless quantities, the speed and the compactness. The dimensionless speed, v/c, gives us an insight into how significant relativistic effects are on the system. The compactness, $\frac{GM}{c^2R}$, where M denotes the total mass of the system and R the orbital separation, shows the strength of the gravitational potential relative to the rest energy of a test particle.

Even the most relativistic binary pulsar, PSR J0737-3039 [14, 15], is still in the relatively slowly varying, weak-field regime from a GR point of view, with a compactness of $GM/(c^2R) \simeq 4.4 \times 10^{-6}$, and a typical speed of $v/c \simeq 2 \times 10^{-3}$. As a reference, the compactness at the Sun's surface is of the order of 10^{-6} and the typical speed of Mercury is of the order of 10^{-4}. It is thus evident that even the binary pulsar systems, which are currently providing the most stringent tests of GR, are still mildly relativistic in nature.

6.4 Gravitational Waves

The *direct* detection of GWs with interferometric detectors such as LIGO and Virgo will provide more stringent tests of GR than binary pulsar measurements. As a comparison, for CBC systems, in the limit of a test particle around a non-spinning BH, the last stable orbit [16] occurs at a separation of $R = 6GM/c^2$, where $GM/(c^2R) = 1/6$ and $v/c = 1/\sqrt{6}$. This constitutes the genuine strong-field, dynamical regime of GR, which, in the foreseeable future, will only be directly accessible by means of GW detectors.

6.4.1 Literature Overview

Indeed, an important goal of direct detections of GWs is the confirmation or falsification of GR in the truly strong-field and dynamical regime. Even as much as a single detection of GWs can make countless experiments that were in accordance with GR redundant. There exists a rich literature in which a number of tests were proposed that can be performed on the direct detection of GWs. What follows is a brief overview of these tests.

6.4.1.1 Gravitational-Wave Properties

One way to test GR is to look at the properties of GWs. For example, GWs propagate at the speed of light. If a GW event is measured with an associated electromagnetic event, one can compare their arrival times. If the source distance is large enough, even a small fractional difference in the speed at which these two types of waves propagate, could lead to a measurable mismatch in their arrival times. Theories such as the Rosen bi-metric theory [17], where GWs follow geodesics of a flat metric and light follows geodesics of a curved metric, and massive graviton theories [18, 19] would both predict such a difference in the arrival times between light and GWs.

Another property of GWs in GR is that they can be described by only two polarisations, as shown in Sect. 1.3. In alternative theories, this is not necessarily true. For example, the Brans-Dicke theory [20] predicts a monopole polarisation, which manifests itself as a so-called breathing mode [21–25]. When a wave arrives perpendicular to the plane of circularly aligned test masses, such a mode would cause the entire ring of test particles to expand uniformly and is therefore invisible to a single interferometric detector, *cf.* Sects. 1.4.3 and 3.1.1. However, a network of detectors does provide the opportunity to measure such a polarisation [26, 27]. A single measurement of the monopole polarisation would also provide evidence that GR may not be the correct theory.

Additionally, one can test the objects that GR predicts. Most notable of such objects are BHs. As the interaction of non-charged BHs is a pure spacetime phenomenon, this provides an excellent way to study the predictions made by GR. Measurements of the BH properties, such as its mass and spin, can empirically test, for example, Penrose's cosmic censorship conjecture [28, 29] and the BH uniqueness theorem [29–35].

6.4.1.2 Direct Search for an Alternative Theory

Besides the searches for unpredicted phenomena, studies have targeted the evolution of binary systems. These tests have focussed on modelling the expected GWs from alternative theories, and the ability to measure the difference between these and the predictions made by GR.

Possible deviations from GR that have been considered in the context of CBC systems include scalar-tensor theories [36–41], a varying Newton constant [42], modified dispersion relation theories, usually referred to in literature as 'massive graviton' models[1] [40, 41, 44–49], and parity violating theories [50–53]. The (rather few) *specific* alternative theories of gravity that have been considered in the context of ground-based GW detectors—essentially scalar-tensor and 'massive graviton' theories—happen to be hard to constrain much further with GW observations. However, GR may be violated in some other manner, including a way that is yet to be

[1] The designations 'massive gravity' and 'massive graviton' originate from Blanchet et al. [43] where only the effect of a modified dispersion relation, or a wavelength dependent propagation speed has been taken into account. While it is attractive to ascribe such a modification to a graviton mass, a modification of the dispersion relation can be a more general effect, and moreover, endowing the graviton with a mass introduces additional deviations from GR than a mere modified dispersion relation. See e.g. the original work by Van Dam and Veltman [18] and the recent work by Rham et al. [19] for a thorough discussion of the issues related to massive gravity models.

envisaged. This makes it imperative to develop methods that can search for *generic* deviations from GR.

6.4.1.3 Generic Searches

Recently, more generic tests of GR through the measurement of GWs have been proposed. One of such tests is due to Arun et al. and exploits the fact that, at least for binaries where neither component has spin, all coefficients ψ_i in the PN expansion of the inspiral phase, given by Eq. (2.89), depend only on the component masses. In that case, only two of the ψ_i are independent. A consistency test of GR can then be constructed by measuring these phase coefficients as extra independent parameters and by comparing any three of them [54–56]. Such a method would be very general, in that one does not have to look for any particular way in which gravity might deviate from GR. Instead, it allows generic failures of GR to be searched for. However, so far its viability was only explored by means of Fisher matrix calculations. Recently, the same philosophy of a consistency check was applied to the quasi-normal modes, and constitute a test of the BH uniqueness theorem [29].

Another proposal is due to Yunes and Pretorius and has been named the PPE formalism [57]. The PPE waveform extends the GR predicted waveforms in the frequency domain with extra amplitude and phase terms. The coefficient and the power of the frequency of the extra terms are left as free parameters. Popular alternative theories are then, up to leading order, cast into the PPE waveform and assigned appropriate values for the PPE parameters [42, 53, 57–63]. A real measurement of the PPE parameters could then reveal the nature of the alternative theory.

In Chap. 3, it was already mentioned that parameter estimation is not the correct framework to test physical theories. Even if the method by Arun et al. or the PPE formalism measures a value of a parameter that conflicts with GR, it does not mean that an alternative theory is more plausible than GR. To answer that question, one needs the framework of model selection. For that, one needs to define two competing hypotheses. One of the hypotheses is the GR hypothesis, which simply states that the observed signal is as predicted by GR. However, the definition for the other hypothesis, in the case of a generic search for GR, is a more challenging task.

6.5 Test Infrastructure for GEneral Relativity (TIGER)

As with any test of a physical law, each of the proposed tests will inevitably have its problems and limitations. For instance, targeted searches will be limited in their testing power as one has to choose the alternative a priori. With the variety of alternatives available, this will prove to be a daunting task.

Generic searches, such as the consistency test by Arun et al., will indeed be a very generic test of the validity of GR, but has limited power to discriminate between different alternatives. Furthermore, the measurement of multiple extra parameters will inevitably diminish one's ability to infer parameters such as the component masses. Especially in the Advanced detector era, where SNRs are expected to be low, poor parameter estimation could be a severe problem. Moreover, it is not immediately clear how one combines information from multiple sources in the case of generic searches such as the consistency test by Arun et al.. To illustrate, suppose we find that for five out of a hundred signal, the signals are inconsistent with GR. Would we assign a five percent probability that GR is violated? What if the five signals show large deviations from GR? Indeed, in the way these generic searches are set up, it is not possible to combine information from multiple sources.

In a search for alternative theories through the measurement of free parameters, one can also be led to an erroneous conclusion if the form of the extension does not exactly model the deviation. This is sometimes referred to as the *fundamental bias* [57]. Unfortunately, no certainty can be obtained unless all functional forms of the extension are tested and compared to one and another. Even if one were to have modelled the deviation in an exact manner, the issue of poor parameter estimation, as explained above, can severely limit the confidence one can assign to a specific alternative theory.

What is called for is a new *inference framework* which avoids the aforementioned problems, while still being exhaustive in its use of the deviations allowed by the model waveforms. Such a framework should comply to the following.

- Be able to test the validity of GR in a generic way.
- Based on hypothesis testing instead of parameter estimation.
- Usable with any waveform approximation or parameterisation of the waveform.
- Avoid explicit and simultaneous estimation of a large set of free parameters, i.e. suitable in the low SNR regime.
- Coherently analyse information from multiple sources.

In what follows, we present TIGER, a framework that incorporates all these restrictions. The method is loosely based on the proposal by Arun et al., in the sense that the framework asks whether coefficients predicted by GR are consistent with the data. However, we start with the application of Bayesian hypothesis testing, in which we try to compare the hypothesis that GR is correct with the hypothesis that GR is incorrect.

In Chap. 7, we will introduce the theoretical framework behind TIGER, and show how one can construct a generic test of GR using Bayesian hypothesis testing. In Chap. 8, we will put the theoretical framework of TIGER to the test through comprehensive simulations of Advanced LIGO/Virgo data. These simulations give insight into the efficiency of TIGER to detect arbitrary deviations from GR. Finally, in Chap. 9, we discuss the results and compare them to existing results in the literature.

References

1. U.J. Le Verrier, Theorie DU mouvement de Mercure. Annales de l'Observatoire de Paris **5**, 1 (1859)
2. A. Einstein, Die Grundlage der allgemeinen Relativitatstheorie. Annalen der Physik **354**, 769–822 (1916)
3. F.W. Dyson, A.S. Eddington, C. Davidson, A determination of the deflection of light by the Sun's gravitational field, from observations made at the total eclipse of May 29, 1919'. English. Philos. Trans. R. Soc. Lond. Ser. A, Containing papers of a mathematical or physical character, **220**, 291–333 (1920)
4. R.V. Pound, G.A. Rebka, Apparent weight of photons. Phys. Rev. Lett. **4**, 337–341 (1960)
5. I.I. Shapiro, Fourth test of general relativity. Phys. Rev. Lett. **13**, 789–791 (1964)
6. B. Bertotti, L. Iess, P. Tortora, A test of general relativity using radio links with the Cassini spacecraft. Nature **425.6956**, 374–376 (2003)
7. F. Zwicky, Nebulae as gravitational lenses. Phys. Rev. **51**, 290–290 (1937)
8. S.B. Lambert, C. Le Poncin-Lafitte, Improved determination of γ by VLBI. A&A **529**(A70), A70 (2011)
9. K. Nordtvedt, The fourth test of general relativity. Rep. Prog. Phys. **45.6**, 631 (1982)
10. F. Hofmann, J. Muller, L. Biskupek, Lunar laser ranging test of the Nordtvedt parameter and a possible variation in the gravitational constant. A&A **522**, L5 (2010)
11. J. Lense, H. Thirring, Uber den Einfluβ der Eigenrotation der Zentralkorperauf die Bewegung der Planeten und Monde nach der Einsteinschen Gravitationstheorie. Physikalische Zeitschrift **19**, 156 (1918)
12. C.W.F. Everitt et al., Gravity probe B: final results of a space experiment to test general relativity. Phys. Rev. Lett. **106**, 221101 (2011)
13. R.A. Hulse, J.H. Taylor, Discovery of a pulsar in a binary system. Astrophys. J. Lett. **195**, L51–L53 (1975)
14. M. Burgay et al., An increased estimate of the merger rate of double neutron stars from observations of a highly relativistic system. Nature **426.6966**, 531–533 (2003)
15. M. Kramer et al., Tests of general relativity from timing the double pulsar. Science **314**(5796), 97–102 (2006)
16. C. Misner, K. Thorne, J. Wheeler, *Gravitation* (WH Freeman & co, San Francisco, 1973)
17. N. Rosen, A bi-metric theory of gravitation. Gen. Relativ. Gravit. **4**, 435–447 (1973)
18. H. van Dam, M. Veltman, Massive and mass-less Yang-Mills and gravitational fields. Nucl. Phys. B **22.2**, 397–411 (1970)
19. C. de Rham, G. Gabadadze, A.J. Tolley, Resummation of massive gravity. Phys. Rev. Lett. **106.23**(231101), 231101 (2011)
20. C. Brans, R.H. Dicke, Mach's principle and a relativistic theory of gravitation. Phys. Rev. **124**, 925–935 (1961)
21. R.V. Wagoner, D. Kalligas, Scalar-tensor theories and gravitational radiation. in *Relativistic Gravitation and Gravitational Radiation*, ed. by J.-A. Marck, J.-P. Lasota (Cambridge University Press, Cambridge, 1997), p. 433
22. M. Brunetti et al., Gravitational wave radiation from compact binary systems in the Jordan-Brans-Dicke theory. Phys. Rev. D **59.4**(044027), 044027 (1999)
23. M. Maggiore, A. Nicolis, Detection strategies for scalar gravitational waves with interferometers and resonant spheres. Phys. Rev. D **62.2**(024004), 024004 (2000)
24. M. Gasperini, On the response of gravitational antennas to dilatonic waves. Phys. Lett. B **470**, 67–72 (1999)
25. L. Wen, B.F. Schutz, Coherent network detection of gravitational waves: the redundancy veto. Class. Quantum Gravity **22**, 1321 (2005)
26. K. Hayama, A. Nishizawa, Model-independent test of gravity with a network of ground-based gravitational-wave detectors. Phys. Rev. D **87.6**(062003), 062003 (2013)

27. K. Chatziioannou, N. Yunes, N. Cornish, Model-independent test of general relativity: an extended post-Einsteinian framework with complete polarization content. Phys. Rev. D **86.2**(022004), 022004 (2012)
28. C. Van Den Broeck, A.S. Sengupta, Binary black hole spectroscopy. Class. Quantum Gravity **24.5**, 1089 (2007)
29. I. Kamaretsos et al., Black-hole hair loss: learning about binary progenitors from ringdown signals. Phys. Rev. D **85.2**(024018), 024018 (2012)
30. F.D. Ryan, Gravitational waves from the inspiral of a compact object into a massive, axisymmetric body with arbitrary multipole moments. Phys. Rev. D **52**, 5707–5718 (1995)
31. E. Poisson, Measuring black-hole parameters and testing general relativity using gravitational-wave data from space-based interferometers. Phys. Rev. D **54**, 5939–5953 (1996)
32. E. Berti, V. Cardoso, C.M. Will, Gravitational-wave spectroscopy of massive black holes with the spaceinterferometer LISA. Phys. Rev. D **73**, 064030 (2006)
33. S.A. Hughes, (Sort of) Testing relativity with extreme mass ratio inspirals. in *AIP Conference Proceedings* ed. by S.M. Merkowitz, J.C. Livas, vol. 873.1 (2006), pp. 233–240
34. L. Barack, C. Cutler, Using LISA extreme-mass-ratio inspiral sources to test off-Kerr deviationsin the geometry of massive black holes. Phys. Rev. D **75**, 042003 (2007)
35. J.R. Gair, C. Li, I. Mandel, Observable properties of orbits in exact bumpy spacetimes. Phys. Rev. D **77**, 024035 (2008)
36. C.M. Will, Testing scalar-tensor gravity with gravitational-wave observations of inspiralling compact binaries. Phys. Rev. D **50**, 6058–6067 (1994)
37. T. Damour, G. Esposito-Farése, Gravitational-wave versus binary-pulsar tests of strong-field gravity. Phys. Rev. D **58**, 042001 (1998)
38. P.D. Scharre, C.M. Will, Testing scalar-tensor gravity using space gravitational wave interferometers. Phys. Rev. D **65**, 042002 (2002)
39. C.M. Will, N. Yunes, Testing alternative theories of gravity using LISA. Class. Quantum Gravity **21.18**, 4367 (2004)
40. E. Berti, A. Buonanno, C.M. Will, Estimating spinning binary parameters and testing alternative theories of gravity with LISA. Phys. Rev. D **71**, 084025 (2005)
41. E. Berti, A. Buonanno, C.M. Will, Testing general relativity and probing the merger history of massive black holes with LISA. Class. Quantum Gravity **22.18**, S943 (2005)
42. N. Yunes, F. Pretorius, D. Spergel, Constraining the evolutionary history of Newton's constant with gravitational wave observations. Phys. Rev. D **81**, 064018 (2010)
43. L. Blanchet et al., Gravitational-radiation damping of compact binary systems to second post-newtonian order. Phys. Rev. Lett. **74**, 3515–3518 (1995)
44. C.M. Will, Bounding the mass of the graviton using gravitational-wave observations of inspiralling compact binaries. Phys. Rev. D **57**, 2061–2068 (1998)
45. K.G. Arun, C.M. Will, Bounding the mass of the graviton with gravitational waves: effect of higher harmonics in gravitational waveform templates. Class. Quantum Gravity **26.15**, 155002 (2009)
46. A. Stavridis, C.M. Will, Bounding the mass of the graviton with gravitational waves: effect of spin precessions in massive black hole binaries. Phys. Rev. D **80**, 044002 (2009)
47. D. Keppel, P. Ajith, Constraining the mass of the graviton using coalescing black-hole binaries. Phys. Rev. D **82**, 122001 (2010)
48. W. Del Pozzo, J. Veitch, A. Vecchio, Testing general relativity using Bayesian model selection: applicationsto observations of gravitational waves from compact binary systems. Phys. Rev. D **83**, 082002 (2011)
49. E. Berti, J. Gair, A. Sesana, Graviton mass bounds from space-based gravitational wave observations of massive black hole populations. Phys. Rev. D **84.10**(101501), 101501 (2011)
50. S. Alexander, L.S. Finn, N. Yunes, Gravitational-wave probe of effective quantum gravity. Phys. Rev. D **78**, 066005 (2008)
51. N. Yunes, L.S. Finn, Constraining effective quantum gravity with LISA. J. Phys. Conf. Ser. **154.1** (2009)

52. C.F. Sopuerta, N. Yunes, Extreme- and intermediate-mass ratio inspirals in dynamical Chern-Simons modified gravity. Phys. Rev. D **80**, 064006 (2009)
53. N. Yunes, R. O'Shaughnessy, B.J. Owen, S. Alexander, Testing gravitational parity violation with coincident gravitational waves and short gamma ray bursts. Phys. Rev. D **82**, 064017 (2010)
54. K.G. Arun et al., Testing post-Newtonian theory with gravitational wave observations. Class. Quantum Gravity **23.9**, L37 (2006)
55. K.G. Arun et al., Probing the nonlinear structure of general relativity with blackhole binaries. Phys. Rev. D **74**, 024006 (2006)
56. C.K. Mishra et al., Parametrized tests of Post-Newtonian theory using advanced LIGO and Einstein telescope. Phys. Rev. D **82**, 064010 (2010)
57. N. Yunes, F. Pretorius, Fundamental theoretical bias in gravitational wave astrophysics and the parametrized post-Einsteinian framework. Phys. Rev. D **80**, 122003 (2009)
58. L.C. Stein, N. Yunes, Effective gravitational wave stress-energy tensor in alternative theories of gravity. Phys. Rev. D **83**, 064038 (2010)
59. N. Yunes, S.A. Hughes, Binary pulsar constraints on the parametrized post-Einsteinian framework. Phys. Rev. D **82**, 082002 (2010)
60. N. Cornish et al., Gravitational wave tests of general relativity with the parameterized post-Einsteinian framework. Phys. Rev. D **84.6**(062003), 062003 (2011)
61. S. Mirshekari, N. Yunes, C.M. Will, Constraining Lorentz-violating, modified dispersion relations with gravitational waves. Phys. Rev. D **85.2**(024041), 024041 (2012)
62. K. Yagi et al., Post-Newtonian, quasicircular binary inspirals in quadratic modified gravity. Phys. Rev. D **85.6**(064022), 064022 (2012)
63. N. Yunes, L.C. Stein, Non-spinning black holes in alternative theories of gravity. Phys. Rev. D **83**, 104002 (2011)

Chapter 7
Test Infrastructure for GEneral Relativity (TIGER)

We will now proceed to introduce TIGER. As was mentioned before, TIGER seeks to test GR by finding the answer to the question "is the measured waveform consistent with GR?". In Chap. 4, we have seen that we need to apply the framework of hypothesis testing in order to answer this question.

As the GW phase is predominantly given by twice the orbital phase of the binary, *cf.* Sects. 1.6 and 2.6, it is particularly sensitive to the underlying laws that govern the dynamics of binary systems. This suggests that if GR were to be incorrect, imprints of the alternative theory can be found in the phase of the GW. As it turns out, the advanced detectors are not sensitive to the subdominant contributions to the amplitude [1, 2]. Without loss of generality, we will therefore consider the phase to be the only possible location of a deviation. However, it is important to bear in mind that TIGER can be applied to any parameterisable part of the waveform e.g. the amplitude in Eq. (2.72), the orbital energy in Eq. (2.65) or luminosity Eq. (2.66). Indeed, an exhaustive test ought to consider deviations in all measurable aspects of the GW.

7.1 Defining the Odds Ratio

The central quantity in hypothesis testing is the *odds ratio*, given by Eq. (4.32). In order to evaluate the odds ratio, we need to define the hypotheses that are to be compared. The two hypotheses we would like to consider are "the waveform has a functional dependence as predicted by GR", denoted by $\mathcal{H}_{\mathrm{GR}}$, and "the waveform does not have a functional dependence as predicted by GR", denoted by $\mathcal{H}_{\mathrm{modGR}}$. The relevant odds ratio is then given by

$$O_{\mathrm{GR}}^{\mathrm{modGR}} \equiv \frac{P(\mathcal{H}_{\mathrm{modGR}}|d, \mathrm{I})}{P(\mathcal{H}_{\mathrm{GR}}|d, \mathrm{I})}. \tag{7.1}$$

T.G.F. Li, *Extracting Physics from Gravitational Waves*,
Springer Theses, DOI 10.1007/978-3-319-19273-4_7

To compare these hypotheses, it will be sufficient to look for a *limited* set of possible deviations. The set of deviations will depend on the waveform approximation being applied to the method. For this work, we consider the waveforms of the PN approximation in Sect. 2.6 and in particular, the `TaylorF2` waveform, described in Sect. 2.6.6. This waveform is used because of its computational speed as well as its agreement with time domain waveforms for BNS systems [3]. Therefore, we take the set of deviations only to be within the known phase coefficients $\{\psi_0, \psi_1, \ldots, \psi_N\}$, given by Eq. (2.89). If the set of phase coefficients has N members, there are $2^N - 1$ ways in which the deviation can occur, corresponding to all possible subsets of the set of phase coefficients.

Before we define and describe the hypotheses $\mathcal{H}_{\rm GR}$ and $\mathcal{H}_{\rm modGR}$, it is instructive to first define the so-called *sub-hypotheses*, denoted by $H_{i_1 i_2 \ldots i_k}$, which are related to the possible deviations in the phase coefficients $\{\psi_0, \psi_1, \ldots, \psi_N\}$, to be

> $H_{i_1 i_2 \ldots i_k}$ is the hypothesis that the phasing coefficients $\psi_{i_1}, \ldots, \psi_{i_k}$ *do not* have the functional dependence on the system parameters as predicted by GR, but all other coefficients ψ_j, $j \notin \{i_1, i_2, \ldots, i_k\}$ *do* have the dependence as predicted by GR.

Thus, for example, H_{12} is the hypothesis that ψ_1 and ψ_2 do not have the functional dependence on the system parameters as predicted by GR, but all other coefficients do. As spins are considered to be negligible for NSs [4], we consider the non-spinning `TaylorF2` waveform, for which the phase coefficients depend only on the two mass parameters. It is also important to note that, by definition of the sub-hypotheses $H_{i_1 i_2 \ldots i_k}$, they are *mutually, logically disjoint*, i.e. $H_{i_1 i_2 \ldots i_k} \wedge H_{j_1 j_2 \ldots j_l}$ is always false for $\{i_1, i_2, \ldots, i_k\} \neq \{j_1, j_2, \ldots, j_l\}$. For example, $H_1 \wedge H_2$ is always false, as H_1 assumes that ψ_2 is in accordance with GR, whereas H_2 assumes that it is not. This will prove to be useful when the main hypotheses are split up into sub-hypotheses.

7.1.1 Defining the GR Hypothesis

Because the two body problem has no analytical solution in GR, GWs can only be computed in an approximate fashion. The approximation can be in the form of a perturbative series or the approximation can be in the finite step size when solving the differential equations numerically. A true test of GR would therefore be difficult to implement in practice. However, one can still test GR to the precision (the PN order in this case) that is currently known. With this restriction in mind, the GR hypothesis, $\mathcal{H}_{\rm GR}$, can be defined as

> $\mathcal{H}_{\rm GR}$ is the hypothesis that the GW has a functional form and coefficients that are exactly as predicted by the chosen approximation to GR.

It should be noted that this definition of $\mathcal{H}_{\rm GR}$ is only valid within the scope of a *specific* waveform approximation. In the case of the PN formalism, $\mathcal{H}_{\rm GR}$ is the hypothesis that the waveform has the functional form as described in e.g. Eq. (2.86).

For BNS systems, we do not expect such a truncation to have a significant effect, as the TaylorF2 waveform family (known up to the 3.5PN order in phase) is known to be in good agreement with numerical simulations [3].

7.1.2 Defining the modGR Hypothesis

Because deviations from GR can be parameterised in an infinite number of ways, testing a hypothesis of all possible deviations is not feasible. Instead, we ask the question whether or not a GW signal adheres to the hypothesis $\mathcal{H}_{\mathrm{GR}}$, defined above. With this in mind, we define the hypothesis $\mathcal{H}_{\mathrm{modGR}}$ to be

> $\mathcal{H}_{\mathrm{modGR}}$ is the hypothesis that one or more of the phase coefficients $\{\psi_0, \psi_1, \ldots, \psi_N\}$ in the waveform do not agree with the prediction made by GR.

However, the hypothesis $\mathcal{H}_{\mathrm{modGR}}$ stated in this form cannot be tested because it is not a generative model. In other words, we do not know how to generate a waveform for which we do not know in which of the phase coefficients the deviations are located. Instead, we can express the hypothesis $\mathcal{H}_{\mathrm{modGR}}$ in terms of the sub-hypotheses $H_{i_1 i_2 \ldots i_k}$, defined on p. 106, and write

$$\mathcal{H}_{\mathrm{modGR}} = \bigvee_{i_1 < i_2 < \ldots < i_k ; k \leq N} H_{i_1 i_2 \ldots i_k}. \tag{7.2}$$

Therefore, for a signal to be inconsistent with GR, we require that one or more phase coefficients deviate from GR. In terms of hypotheses, we are thus interested in the disjunction of the sub-hypotheses $H_{i_1 i_2 \ldots i_k}$.

To date, the TaylorF2 phase has ten known phase coefficients, namely $\psi_0, \ldots, \psi_7$ and two additional coefficients $\psi_5^{(l)}$ and $\psi_6^{(l)}$, associated with logarithmic contributions [see Eq. (2.89)]. In this book, we will not use ψ_0 as a variable coefficient, since it has been tested by binary pulsars measurements (see Lorimer [5] for an overview). Even so, if one were to consider all the subsets of the set of remaining coefficients, one would have to take into account the $2^9 - 1 = 511$ ways in which a deviation can occur. Apart from this being computationally demanding, we do not expect advanced detectors to be sensitive to the highest-order coefficients. Therefore, it makes sense to limit oneself to all the subsets of the *testing coefficients*

$$\{\psi_1, \psi_2, \ldots, \psi_{N_T}\}, \tag{7.3}$$

where N_T is the number of these testing coefficients one chooses to consider. We thus allow one or more of the coefficients in the set $\{\psi_1, \psi_2 \ldots, \psi_{N_T}\}$ to vary freely, instead of following the functional dependence on the component masses as predicted by GR. The choice of N_T will in part be influenced by the required generality of the test, measurability of phase coefficients and computational limitations.

7.1.3 Odds Ratio for a Single Source: A 2 Testing Coefficient Example

Suppose that only two coefficients, ψ_1 and ψ_2, are chosen to be our testing coefficients. Then, the hypothesis $\mathcal{H}_{\mathrm{modGR}}$ in Eq. (7.2) can be written as

$$\mathcal{H}_{\mathrm{modGR}} = H_1 \vee H_2 \vee H_{12}. \tag{7.4}$$

In this example, the odds ratio of interest, which is defined in Eq. (7.1), can be written as

$${}^{(2)}O_{\mathrm{GR}}^{\mathrm{modGR}} = \frac{P(H_1 \vee H_2 \vee H_{12}|d, \mathrm{I})}{P(\mathcal{H}_{\mathrm{GR}}|d, \mathrm{I})}, \tag{7.5}$$

where the superscript (2) reminds us that only two of the phase coefficients are being used as testing coefficients.

An important observation is that the sub-hypotheses H_1, H_2, H_{12} are *mutually, logically disjoint*, i.e. the conjunction of any two of them is always false. Indeed, in H_1, ψ_2 takes the value predicted by GR, but in H_2 it differs from the GR value, as it does in H_{12}. Similarly, in H_2, ψ_1 takes the GR value, but in H_1 it differs from the GR value, and the same in H_{12}. This means that the odds ratio is simply given by

$${}^{(2)}O_{\mathrm{GR}}^{\mathrm{modGR}} = \frac{P(H_1|d, \mathrm{I})}{P(\mathcal{H}_{\mathrm{GR}}|d, \mathrm{I})} + \frac{P(H_2|d, \mathrm{I})}{P(\mathcal{H}_{\mathrm{GR}}|d, \mathrm{I})} + \frac{P(H_{12}|d, \mathrm{I})}{P(\mathcal{H}_{\mathrm{GR}}|d, \mathrm{I})}. \tag{7.6}$$

Using Bayes' theorem in Eq. (4.31), this can be written as

$${}^{(2)}O_{\mathrm{GR}}^{\mathrm{modGR}} = \frac{P(H_1|\mathrm{I})}{P(\mathcal{H}_{\mathrm{GR}}|\mathrm{I})} B_{\mathrm{GR}}^{1} + \frac{P(H_2|\mathrm{I})}{P(\mathcal{H}_{\mathrm{GR}}|\mathrm{I})} B_{\mathrm{GR}}^{2} + \frac{P(H_{12}|\mathrm{I})}{P(\mathcal{H}_{\mathrm{GR}}|\mathrm{I})} B_{\mathrm{GR}}^{12}. \tag{7.7}$$

Here, B_{GR}^{1}, B_{GR}^{2}, B_{GR}^{12} are the Bayes factors of the sub-hypotheses against GR [cf. Eq. (4.33)], given by

$$B_{\mathrm{GR}}^{1} = \frac{P(d|H_1, \mathrm{I})}{P(d|\mathcal{H}_{\mathrm{GR}}, \mathrm{I})}, \qquad B_{\mathrm{GR}}^{2} = \frac{P(d|H_2, \mathrm{I})}{P(d|\mathcal{H}_{\mathrm{GR}}, \mathrm{I})}, \qquad B_{\mathrm{GR}}^{12} = \frac{P(d|H_{12}, \mathrm{I})}{P(d|\mathcal{H}_{\mathrm{GR}}, \mathrm{I})}, \tag{7.8}$$

and $P(H_1|\mathrm{I})/P(\mathcal{H}_{\mathrm{GR}}|\mathrm{I})$, $P(H_2|\mathrm{I})/P(\mathcal{H}_{\mathrm{GR}}|\mathrm{I})$ and $P(H_{12}|\mathrm{I})/P(\mathcal{H}_{\mathrm{GR}}|\mathrm{I})$ are ratios of prior odds.

Now, upon calculating the Bayes factors for each sub-hypothesis, we would like to combine these measurements into an overall odds ratio between the GR hypothesis and *any* of the competing hypotheses in Eq. (7.7). In order to do this, we must specify the prior odds for each sub-hypothesis against GR, i.e. $P(H_1|\mathrm{I})/P(\mathcal{H}_{\mathrm{GR}}|\mathrm{I})$, $P(H_2|\mathrm{I})/P(\mathcal{H}_{\mathrm{GR}}|\mathrm{I})$ and $P(H_{12}|\mathrm{I})/P(\mathcal{H}_{\mathrm{GR}}|\mathrm{I})$. Here one might want to let oneself be guided by, for example, the expectation that a violation of GR will likely occur at higher PN order, and give more weight to H_2 and H_{12}. Or, if one expects a deviation to happen only in a particular phase coefficient (such as ψ_2 in the case of 'massive gravity'), one may want to downweigh the most inclusive sub-hypothesis, in this example H_{12}. In reality, we will not know beforehand what form a violation will take. In particular, it could affect all the PN coefficients. For the purposes of this analysis, we invoke the principle of indifference among all sub-hypotheses, i.e. taking no one to be preferable to any other. This imposes the condition that the prior odds of each against GR are equal. When combining the Bayes factors into the odds ratio, we therefore assume

$$\frac{P(H_1|\mathrm{I})}{P(\mathcal{H}_{\mathrm{GR}}|\mathrm{I})} = \frac{P(H_2|\mathrm{I})}{P(\mathcal{H}_{\mathrm{GR}}|\mathrm{I})} = \frac{P(H_{12}|\mathrm{I})}{P(\mathcal{H}_{\mathrm{GR}}|\mathrm{I})}. \tag{7.9}$$

By invoking the principle of indifference, we can factorise out the prior odds ratios. However, we are still left to specify the values for these prior odds ratios. Equivalently, we can do this by assigning a value to the *overall* prior odds ratio, given by

$$\frac{P(\mathcal{H}_{\mathrm{modGR}}|\mathrm{I})}{P(\mathcal{H}_{\mathrm{GR}}|\mathrm{I})} = \frac{P(H_1 \vee H_2 \vee H_{12}|\mathrm{I})}{P(\mathcal{H}_{\mathrm{GR}}|\mathrm{I})} = \alpha, \tag{7.10}$$

where we do not yet specify the value for α. As it turns out, it will end up being an overall scaling of the odds ratio. This, together with Eq. (7.9) and the logical disjointness of the sub-hypotheses H_1, H_2 and H_{12} implies that

$$\frac{P(H_1|\mathrm{I})}{P(\mathcal{H}_{\mathrm{GR}}|\mathrm{I})} = \frac{P(H_2|\mathrm{I})}{P(\mathcal{H}_{\mathrm{GR}}|\mathrm{I})} = \frac{P(H_{12}|\mathrm{I})}{P(\mathcal{H}_{\mathrm{GR}}|\mathrm{I})} = \frac{\alpha}{3}. \tag{7.11}$$

The final expression for the odds ratio for a modification of GR versus GR, in the case where two phase coefficients are used as our testing coefficients, is given by

$$^{(2)}O_{\mathrm{GR}}^{\mathrm{modGR}} = \frac{\alpha}{3}\left[B_{\mathrm{GR}}^{1} + B_{\mathrm{GR}}^{2} + B_{\mathrm{GR}}^{12}\right]. \tag{7.12}$$

Therefore, up to the overall scaling factor α, the odds ratio is simply the *average* of the Bayes factors.

7.1.4 Odds Ratio for a Single Source: The General Case

Next, we can generalise the expression of the odds ratio for N_T testing coefficients as

$$\begin{aligned} ^{(N_T)}O_{\mathrm{GR}}^{\mathrm{modGR}} &= \frac{P(\mathcal{H}_{\mathrm{modGR}}|d,\mathrm{I})}{P(\mathcal{H}_{\mathrm{GR}}|d,\mathrm{I})} \\ &= \frac{P(\bigvee_{i_1<i_2<\ldots<i_k;k\leq N_T} H_{i_1 i_2 \ldots i_k}|d,\mathrm{I})}{P(\mathcal{H}_{\mathrm{GR}}|d,\mathrm{I})}. \end{aligned} \tag{7.13}$$

By using the fact that the sub-hypotheses are mutually, logically disjoint, one can write

$$P(\bigvee_{i_1<i_2<\ldots<i_k;k\leq N_T} H_{i_1 i_2\ldots i_k}|d,\mathrm{I}) = \sum_{k=1}^{N_T}\sum_{i_1<i_2<\ldots<i_k} P(H_{i_1 i_2\ldots i_k}|d,\mathrm{I}). \tag{7.14}$$

Similar to Eq. (4.32), we can rewrite the odds ratio as

$$^{(N_T)}O_{\mathrm{GR}}^{\mathrm{modGR}} = \sum_{k=1}^{N_T}\sum_{i_1<i_2<\ldots<i_k} \frac{P(H_{i_1 i_2\ldots i_k}|\mathrm{I})}{P(\mathcal{H}_{\mathrm{GR}}|\mathrm{I})} B_{\mathrm{GR}}^{i_1 i_2\ldots i_k}, \tag{7.15}$$

where $P(H_{i_1 i_2\ldots i_k}|\mathrm{I})/P(\mathcal{H}_{\mathrm{GR}}|\mathrm{I})$ denote the prior odds ratios and

$$B_{\mathrm{GR}}^{i_1 i_2\ldots i_k} = \frac{P(d|H_{i_1 i_2\ldots i_k},\mathrm{I})}{P(d|\mathcal{H}_{\mathrm{GR}},\mathrm{I})} \tag{7.16}$$

are the Bayes factors for the sub-hypotheses against GR. At this point, one has to set the values for the relative prior probabilities, $P(H_{i_1 i_2\ldots i_k}|\mathrm{I})/P(\mathcal{H}_{\mathrm{GR}}|\mathrm{I})$. Once again, we invoke the principle of indifference and assign to each an equal weight, i.e.

$$\frac{P(H_{i_1 i_2\ldots i_k}|\mathrm{I})}{P(\mathcal{H}_{\mathrm{GR}}|\mathrm{I})} = \frac{P(H_{j_1 j_2\ldots j_l}|\mathrm{I})}{P(\mathcal{H}_{\mathrm{GR}}|\mathrm{I})} \quad \text{for any } k, l \leq N_T. \tag{7.17}$$

We can further specify the prior odds by considering the overall prior odds,

$$\begin{aligned}\frac{P(\mathcal{H}_{\mathrm{modGR}}|\mathrm{I})}{P(\mathcal{H}_{\mathrm{GR}}|\mathrm{I})} &= \frac{P(\bigvee_{i_1<i_2<\ldots<i_k;k\leq N_T} H_{i_1 i_2\ldots i_k}|\mathrm{I})}{P(\mathcal{H}_{\mathrm{GR}}|\mathrm{I})}\\ &= \sum_{k=1}^{N_T}\sum_{i_1<i_2<\ldots<i_k} \frac{P(H_{i_1 i_2\ldots i_k}|\mathrm{I})}{P(\mathcal{H}_{\mathrm{GR}}|\mathrm{I})}.\end{aligned} \tag{7.18}$$

The overall prior odds describes the relative prior belief between $\mathcal{H}_{\mathrm{GR}}$ and $\mathcal{H}_{\mathrm{modGR}}$, and can be seen as an overall normalisation factor. As the choice of this quantity depends in essence on the definition of the background information, we write

$$\frac{P(\mathcal{H}_{\mathrm{modGR}}|\mathrm{I})}{P(\mathcal{H}_{\mathrm{GR}}|\mathrm{I})} = \alpha, \tag{7.19}$$

and leave the assignment of the overall prior odds ratio to the reader's choice. As will become apparent, the overall relative prior will end up being an overall scaling of the odds ratio, and is unimportant when one considers the method in light of signals embedded in noise.

The equalities (7.18) and (7.19), together with (7.17), imply

$$\frac{P(H_{i_1 i_2 \ldots i_k}|\mathrm{I})}{P(\mathcal{H}_{\mathrm{GR}}|\mathrm{I})} = \frac{\alpha}{2^{N_T} - 1}. \tag{7.20}$$

In terms of the sub-hypotheses $H_{i_1 i_2 \ldots i_k}$, the odds ratio can be written as

$$^{(N_T)}\mathcal{O}^{\mathrm{modGR}}_{\mathrm{GR}} = \frac{\alpha}{2^{N_T} - 1} \sum_{k=1}^{N_T} \sum_{i_1 < i_2 < \ldots < i_k} B^{i_1 i_2 \ldots i_k}_{\mathrm{GR}}. \tag{7.21}$$

Up to an overall prefactor, the odds ratio is thus a straightforward *average* of the Bayes factors from the $2^{N_T} - 1$ individual sub-hypotheses.

7.1.5 Odds Ratio for Multiple Sources

Although the detection rate for CBC sources is still rather uncertain, we expect advanced detectors to detect several events per year [6]. It is therefore important to take advantage of multiple detections to provide tighter constraints on the validity of GR. Bayesian hypothesis testing provides a natural extension of the odds ratio from a quantity stemming from a single measurement to a quantity which takes a catalogue of sources into consideration.

Consider $\mathcal{N}$ independent data sets, denoted by $\{d_1, d_2, \ldots, d_{\mathcal{N}}\}$, corresponding to $\mathcal{N}$ independent GW events. We do not assume that deviations from GR are necessarily the same between events, but rather that they can vary from source to source. For example, the deviation from GR could depend on the masses or even on unknown additional charges in alternative theories. One can write down an odds ratio for a catalogue of sources as

$$\begin{aligned} ^{(N_T)}\mathcal{O}^{\mathrm{modGR}}_{\mathrm{GR}} &= \frac{P(\mathcal{H}_{\mathrm{modGR}}|d_1, \ldots, d_{\mathcal{N}}, \mathrm{I})}{P(\mathcal{H}_{\mathrm{GR}}|d_1, \ldots, d_{\mathcal{N}}, \mathrm{I})} \\ &= \frac{\sum_{k=1}^{N_T} \sum_{i_1 < i_2 < \ldots < i_k} P(H_{i_1 i_2 \ldots i_k}|d_1, \ldots, d_{\mathcal{N}}, \mathrm{I})}{P(\mathcal{H}_{\mathrm{GR}}|d_1, \ldots, d_{\mathcal{N}}, \mathrm{I})} \\ &= \sum_{k=1}^{N_T} \sum_{i_1 < i_2 < \ldots < i_k} \frac{P(H_{i_1 i_2 \ldots i_k}|\mathrm{I})}{P(\mathcal{H}_{\mathrm{GR}}|\mathrm{I})} \, {}^{(\mathrm{cat})}B^{i_1 i_2 \ldots i_k}_{\mathrm{GR}}, \end{aligned} \tag{7.22}$$

where the catalogue Bayes factors are given by

$$^{(\mathrm{cat})}B^{i_1 i_2 \ldots i_k}_{\mathrm{GR}} = \frac{P(d_1, \ldots, d_{\mathcal{N}}|H_{i_1 i_2 \ldots i_k}, \mathrm{I})}{P(d_1, \ldots, d_{\mathcal{N}}|\mathcal{H}_{\mathrm{GR}}, \mathrm{I})}. \tag{7.23}$$

Assuming that the events $d_1, \ldots, d_{\mathcal{N}}$ are all independent, one has

$$\begin{aligned}
{}^{(\mathrm{cat})}B^{i_1 i_2 \ldots i_k}_{\mathrm{GR}} &= \frac{P(d_1, \ldots, d_{\mathcal{N}} | H_{i_1 i_2 \ldots i_k}, \mathrm{I})}{P(d_1, \ldots, d_{\mathcal{N}} | \mathcal{H}_{\mathrm{GR}}, \mathrm{I})} \\
&= \frac{P(d_1 | H_{i_1 i_2 \ldots i_k}, \mathrm{I})}{P(d_1 | \mathcal{H}_{\mathrm{GR}}, \mathrm{I})} \frac{P(d_2, \ldots, d_{\mathcal{N}} | d_1, H_{i_1 i_2 \ldots i_k}, \mathrm{I})}{P(d_2, \ldots, d_{\mathcal{N}} | d_1, \mathcal{H}_{\mathrm{GR}}, \mathrm{I})} \\
&= \frac{P(d_1 | H_{i_1 i_2 \ldots i_k}, \mathrm{I})}{P(d_1 | \mathcal{H}_{\mathrm{GR}}, \mathrm{I})} \frac{P(d_2, \ldots, d_{\mathcal{N}} | H_{i_1 i_2 \ldots i_k}, \mathrm{I})}{P(d_2, \ldots, d_{\mathcal{N}} | \mathcal{H}_{\mathrm{GR}}, \mathrm{I})} \\
&\;\;\vdots \\
&= \prod_{A=1}^{\mathcal{N}} \frac{P(d_A | H_{i_1 i_2 \ldots i_k}, \mathrm{I})}{P(d_A | \mathcal{H}_{\mathrm{GR}}, \mathrm{I})} \\
&= \prod_{A=1}^{\mathcal{N}} {}^{(A)}B^{i_1 i_2 \ldots i_k}_{\mathrm{GR}}, \qquad (7.24)
\end{aligned}$$

where ${}^{(A)}B^{i_1 i_2 \ldots i_k}_{\mathrm{GR}}$ is the Bayes factor for event A.

To evaluate the combined odds ratio of the catalogue, we choose once again to invoke indifference [as in Eq. (7.17)] and set the individual prior odds ratios equal to each other, so that

$$\frac{P(H_{i_1 i_2 \ldots i_k} | \mathrm{I})}{P(\mathcal{H}_{\mathrm{GR}} | \mathrm{I})} = \frac{\alpha}{2^{N_T} - 1}. \qquad (7.25)$$

Together with Eqs. (7.22) and (7.24), this leads to

$${}^{(N_T)}\mathcal{O}^{\mathrm{modGR}}_{\mathrm{GR}} = \frac{\alpha}{2^{N_T} - 1} \sum_{k=1}^{N_T} \sum_{i_1 < i_2 < \ldots < i_k} \prod_{A=1}^{\mathcal{N}} {}^{(A)}B^{i_1 i_2 \ldots i_k}_{\mathrm{GR}}, \qquad (7.26)$$

which, up to an overall prefactor, amounts to taking the average of the catalogue Bayes factors, given in Eq. (7.24).

The expression in Eq. (7.26) shows that the odds ratio for a catalogue of sources can be constructed as a linear combination of the product of the Bayes factors from single sources. As a consequence, one can analyse sources individually without losing the ability to construct the Bayes factor for a catalogue. Furthermore, the odds ratio can be computed as a parallel process, which greatly decreases the computational time.

Alternatively, one may prefer not to make any assumptions about the prior odds ratios $P(H_{i_1 i_2 \ldots i_k} | \mathrm{I}) / P(\mathcal{H}_{\mathrm{GR}} | \mathrm{I})$ at all, and focus on the catalogue Bayes factors ${}^{(\mathrm{cat})}B^{i_1 i_2 \ldots i_k}_{\mathrm{GR}}$ separately and individually. Although only the odds ratio can tell us whether GR or some alternative is preferred, the Bayes factors can be a powerful tool in analysing the performance of TIGER.

It should be pointed out that the seemingly trivial extension of the odds ratio for a single source to one for a catalogue of sources is cardinal. A careful look at the components making up the odds ratio reveals that this quantity is solely dependent on the sub-hypotheses themselves. In other words, the odds ratio has no explicit dependence on the value of any parameters and depends only on the data and the background information.

This, however, is not generally true when a test is performed by measuring coefficients alone. Consider an alternative theory that causes deviations from GR that are mass dependent. Combining posteriors on the measured parameters is then impossible unless one knows the functional dependence of the deviations on the masses, a priori. Only in the case in which the deviation from GR is governed by constants can one combine information from multiple sources. Therefore, a test of GR is more versatile and robust if it compares hypotheses instead of measuring coefficients. To make an even stronger statement: an alternative theory ought not to be interpreted as to be favoured over GR unless the odds ratio shows such indication.

7.2 Quantification of Noisy Measurements

From a theoretical point of view, the data favour the hypothesis $\mathcal{H}_{\text{modGR}}$ over to the hypothesis $\mathcal{H}_{\text{GR}}$ when $\mathcal{O}_{\text{GR}}^{\text{modGR}} > 1$, *cf.* Eqs (7.13) and (7.22). The relative degree of belief in the two hypotheses is encapsulated in the magnitude of the odds ratio. However, in the case of advanced detectors, the signals will be buried deep inside the noise. This introduces the problem that the noise itself can mimic the effects of a deviation from GR that is non-negligible. Hence, we need to develop the tools to correctly account for the presence of noise in our detectors.

7.2.1 False Alarm Probability (FAP)

To gauge the influence of noise in the detectors on our interpretation of the odds ratio, one constructs a distribution of odds ratios from simulated signals, collectively denoted by κ, that are consistent with $\mathcal{H}_{\text{GR}}$, but are also embedded within noise. This is the so-called *background* distribution, $P\left(\ln {}^{(N_T)}\mathcal{O}_{\text{GR}}^{\text{modGR}} | \kappa, \mathcal{H}_{\text{GR}}, \text{I}\right)$, where we have used the logarithm of the odds ratio to account for its dynamical range. In the advanced-detector era, one will only have access to a single catalogue of events and therefore a single measured odds ratio, denoted by $\overline{{}^{(N_T)}\mathcal{O}_{\text{GR}}^{\text{modGR}}}$. This measured odds ratio can be compared to the simulated background distribution to assign a FAP, denoted by β, given by

$$\beta = \int_{\ln \overline{\mathcal{O}_{\text{GR}}^{\text{modGR}}}}^{\infty} P(\ln \mathcal{O}_{\text{GR}}^{\text{modGR}} | \kappa, \mathcal{H}_{\text{GR}}, \text{I}) \, d \ln \mathcal{O}_{\text{GR}}^{\text{modGR}}. \tag{7.27}$$

Conversely, one can also establish a *threshold*, which the odds ratio of a given catalogue must overcome in order that a violation of GR becomes credible. Given a maximum FAP β_{max}, a threshold for the odds ratio, denoted by $\ln \mathcal{O}_\beta$, can be set as

$$\beta_{\text{max}} = \int_{\ln \mathcal{O}_\beta}^{\infty} P(\ln \mathcal{O}_{\text{GR}}^{\text{modGR}} | \kappa, \mathcal{H}_{\text{GR}}, \text{I})\, d \ln \mathcal{O}_{\text{GR}}^{\text{modGR}}. \tag{7.28}$$

A schematic example of the background distribution and FAP is shown in Fig. 7.1. Shown in dotted and filled areas (blue) is background distribution $P\left(\ln \mathcal{O}_{\text{GR}}^{\text{modGR}} | \kappa, \mathcal{H}_{\text{GR}}, \text{I}\right)$. The dashed line (red) represents the measured odds ratio $\ln \overline{{}^{(N_T)}\mathcal{O}_{\text{GR}}^{\text{modGR}}}$ or the threshold $\ln \mathcal{O}_\beta$. The region making up the FAP is the filled region of the background distribution (blue).

Note that the FAP is invariant under the choice of the overall prior odds ratio α, given by Eq. (7.19). Suppose we rescale the overall prior odds ratio α with a factor γ, i.e.

$$\begin{aligned} \alpha \rightarrow \alpha' &= \gamma\alpha \\ \ln \mathcal{O}_{\text{GR}}^{\text{modGR}} \rightarrow \ln \mathcal{O'}_{\text{GR}}^{\text{modGR}} &= \ln \gamma \mathcal{O}_{\text{GR}}^{\text{modGR}} \\ &= \ln \gamma + \ln \mathcal{O}_{\text{GR}}^{\text{modGR}} \\ d \ln \mathcal{O'}_{\text{GR}}^{\text{modGR}} &= d \ln \mathcal{O}_{\text{GR}}^{\text{modGR}}. \end{aligned} \tag{7.29}$$

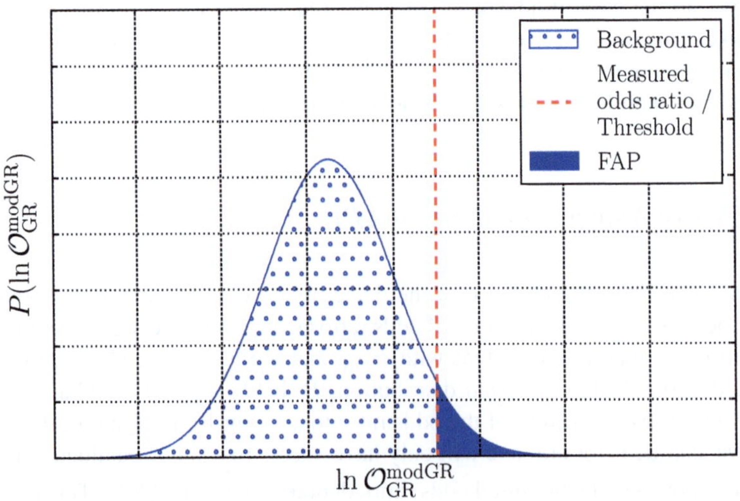

Fig. 7.1 The *dotted* distribution (*blue*) depicts the *background* distribution obtained by analysing a collection κ of signals that satisfy the hypothesis $\mathcal{H}_{\text{GR}}$ and are embedded in detector noise, given by $P\left(\ln \mathcal{O}_{\text{GR}}^{\text{modGR}} | \kappa, \mathcal{H}_{\text{GR}}, \text{I}\right)$. The *dashed line* (*red*) and *shaded area* (*blue*) of the *background* distribution depict the measured odds ratio $\ln \overline{\mathcal{O}_{\text{GR}}^{\text{modGR}}}$ and the associated FAP β, or the threshold $\ln \mathcal{O}_\beta$ for a maximum FAP β_{max}

Now let $P'(\ln \mathcal{O}'^{\text{modGR}}_{\text{GR}}|\kappa, \mathcal{H}_{\text{GR}}, \text{I})$ and $P(\ln \mathcal{O}^{\text{modGR}}_{\text{GR}}|\kappa, \mathcal{H}_{\text{GR}}, \text{I})$ be the distributions of $\ln \mathcal{O}'^{\text{modGR}}_{\text{GR}}$ and $\ln \mathcal{O}^{\text{modGR}}_{\text{GR}}$, respectively. These two distributions are related by

$$P'(\ln \mathcal{O}^{\text{modGR}}_{\text{GR}} + \ln \gamma|\kappa, \mathcal{H}_{\text{GR}}, \text{I}) = P(\ln \mathcal{O}^{\text{modGR}}_{\text{GR}}|\kappa, \mathcal{H}_{\text{GR}}, \text{I}). \tag{7.30}$$

Then, the FAP transforms as

$$\begin{aligned}
\beta \to \beta' &= \int_{\overline{\ln \mathcal{O}'^{\text{modGR}}_{\text{GR}}}}^{\infty} P'(\ln \mathcal{O}'^{\text{modGR}}_{\text{GR}}|\kappa, \mathcal{H}_{\text{GR}}, \text{I})\, d\ln \mathcal{O}'^{\text{modGR}}_{\text{GR}} \\
&= \int_{\overline{\ln \mathcal{O}'^{\text{modGR}}_{\text{GR}}}-\ln\gamma}^{\infty} P'(\ln \mathcal{O}^{\text{modGR}}_{\text{GR}} + \ln\gamma|\kappa, \mathcal{H}_{\text{GR}}, \text{I})\, d\ln \mathcal{O}^{\text{modGR}}_{\text{GR}} \\
&= \int_{\overline{\ln \mathcal{O}^{\text{modGR}}_{\text{GR}}}}^{\infty} P(\ln \mathcal{O}^{\text{modGR}}_{\text{GR}}|\kappa, \mathcal{H}_{\text{GR}}, \text{I})\, d\ln \mathcal{O}^{\text{modGR}}_{\text{GR}} \\
&= \beta
\end{aligned} \tag{7.31}$$

Therefore, the FAP stays invariant under the choice of the overall prior. This is because both the background and the measured odds ratio are only translated and the integral in Eq. (7.27) remains unchanged.

7.2.2 Efficiency in Detecting a Particular Deviation

In the absence of readily available signals, a real catalogue cannot yet be analysed and compared to the background. However, one can still make the assessment of how likely it will be to get a certain odds ratio for a *specific* deviation from GR. In order to do that, one constructs the so-called *foreground*, $P(\ln \mathcal{O}^{\text{modGR}}_{\text{GR}}|\kappa', \mathcal{H}_{\text{alt}}, \text{I})$, by calculating the distribution of odds ratios assuming a collection of signals, denoted by κ', for which some specific alternative theory, denoted by $\mathcal{H}_{\text{alt}}$, is correct. By direct comparison of the foreground to the background, one can infer the probability that a specific alternative theory can cause a deviation that is more significant than the deviations induced by the noise.

We can quantify the probability that a specific alternative theory will be detected with a FAP smaller than a given β_{max}, by means of an *efficiency* ζ, defined as

$$\zeta = \int_{\ln \mathcal{O}_\beta}^{\infty} P(\ln \mathcal{O}^{\text{modGR}}_{\text{GR}}|\kappa', \mathcal{H}_{\text{alt}}, \text{I})\, d\ln \mathcal{O}^{\text{modGR}}_{\text{GR}}, \tag{7.32}$$

where $\ln \mathcal{O}_\beta$ is the threshold defined in Eq. (7.28). Note that the hypothesis $\mathcal{H}_{\text{alt}}$ should not be confused with the hypothesis $\mathcal{H}_{\text{modGR}}$. The hypothesis $\mathcal{H}_{\text{alt}}$ refers to a specific alternative theory, e.g. Brans-Dicke theory, which is tested by the calculation of the foreground. On the other hand, the hypothesis $\mathcal{H}_{\text{modGR}}$ is used for the calculation of the odds ratio in Eq. (7.21) or Eq. (7.26) and refers to the hypothesis

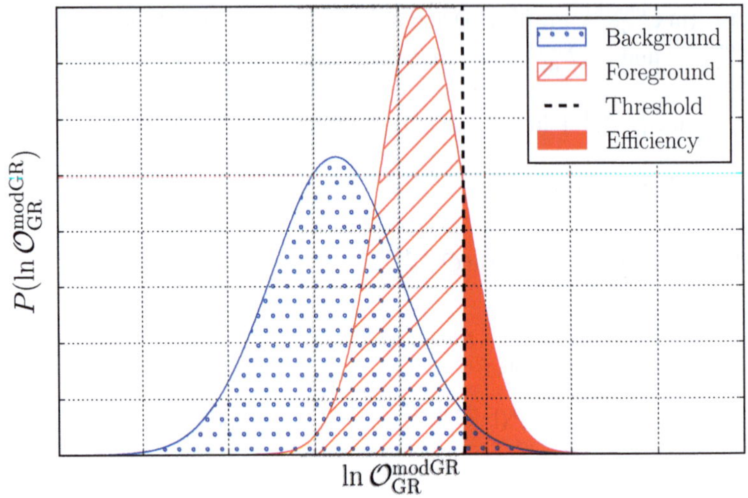

Fig. 7.2 The *blue* distribution depicts the background distribution, given by $P\left(\ln \mathcal{O}_{\rm GR}^{\rm modGR}|\kappa, \mathcal{H}_{\rm GR}, {\rm I}\right)$, obtained by analysing a collection κ of signals that satisfy the $\mathcal{H}_{\rm GR}$ hypothesis and are embedded in detector noise. The *red* distribution is the foreground, given by $P(\mathcal{O}_{\rm GR}^{\rm modGR}|\kappa', \mathcal{H}_{\rm alt}, {\rm I})$, obtained by analysing a collection κ' of signals that follows some alternative theory $\mathcal{H}_{\rm alt}$. The efficiency ζ, shown as the *shaded area* (*red*) of the foreground, is the fraction of foreground above the threshold $\ln \mathcal{O}_\beta$, depicted by the *dashed line* (*black*), from a given maximum FAP $\beta_{\rm max}$

that at least one of the phase coefficients in Eq. (2.89) does not have the functional dependence as predicted by GR.

A schematic example of the efficiency ζ is shown in Fig. 7.2. In this figure, the background is shown as the dotted distribution (blue), the foreground as the dashed distribution (red), and the threshold $\ln \mathcal{O}_\beta$ and efficiency ζ for a given maximum FAP $\beta_{\rm max}$ are depicted as the dashed line (black) and the shaded part (red) of the foreground, respectively.

Just like the FAP, the efficiency is also invariant under the transformation $\alpha \to \alpha' = \gamma\alpha$. Since the FAP stays invariant under this transformation [see Eq. (7.31)], we can deduce that the threshold transforms as

$$\ln \mathcal{O}_\beta \to \ln \mathcal{O}'_\beta = \ln \mathcal{O}_\beta + \ln \gamma. \tag{7.33}$$

Therefore, under the transformation given in Eq. (7.29), the efficiency transforms as

$$\begin{aligned}
\zeta \to \zeta' &= \int_{\ln \mathcal{O}'_\beta}^{\infty} P'(\ln \mathcal{O}'^{\rm modGR}_{\rm GR}|\kappa', \mathcal{H}_{\rm alt}, {\rm I})\, d \ln \mathcal{O}'^{\rm modGR}_{\rm GR} \\
&= \int_{\ln \mathcal{O}'_\beta - \ln \gamma}^{\infty} P'(\ln \mathcal{O}_{\rm GR}^{\rm modGR} + \ln \gamma|\kappa', \mathcal{H}_{\rm alt}, {\rm I})\, d \ln \mathcal{O}_{\rm GR}^{\rm modGR}
\end{aligned}$$

$$= \int_{\ln \mathcal{O}_\beta}^{\infty} P(\ln \mathcal{O}_{\rm GR}^{\rm modGR} | \kappa', \mathcal{H}_{\rm alt}, {\rm I})\, d \ln \mathcal{O}_{\rm GR}^{\rm modGR}$$
$$= \zeta. \tag{7.34}$$

It is evident that, under the transformation shown in Eq. (7.29), both the background and foreground shift by an equal amount, making the value of the overall prior odds ratio unimportant if one is interested in the plausibility of detecting the deviation from a specific alternative theory that is more significant than the background distribution.

7.3 Implementation

A few remarks have to be made regarding the implementation of TIGER. Firstly, we need to explain the implementation of the sub-hypotheses on p. 106. The generative models $H_{i_1 i_2 \ldots i_k}$ state that the phase coefficients $\psi_{i_1}, \ldots, \psi_{i_k}$ do not have the functional dependence on the system parameters as predicted by GR, but all other phase coefficients $\psi_j, j \notin \{i_1, i_2, \ldots, i_k\}$ do. This is treated by allowing the relevant phase coefficients to vary from the GR predicted functional form by adding a freely varying fractional shift. This modification to the phase coefficients ψ_i can be written as

$$\psi_i = \psi_i^{\rm GR}(\mathcal{M}, \eta)\ [1 + \delta\chi_i], \tag{7.35}$$

where $\psi_i^{\rm GR}(\mathcal{M}, \eta)$ is the functional form of ψ_i according to GR [given by Eq. (2.89)], and the dimensionless $\delta\chi_i$ is a fractional shift in ψ_i. However, the 0.5PN phase coefficient, given by ψ_1, is identically zero and cannot be implemented in a similar fashion. Instead, deviations from $\psi_1^{\rm GR}$ are modelled as

$$\psi_1^{\rm GR}(\mathcal{M}, \eta) = 0 \rightarrow \frac{3}{128\eta}(\pi M)^{-4/3}\delta\chi_1, \tag{7.36}$$

and the interpretation of a fractional shift is not adequate. Instead, $\delta\chi_1$ is related to the magnitude of the deviation itself.

Secondly, in order to construct the odds ratio in Eq. (7.21) or Eq. (7.26), we need to calculate the evidence associated to the various sub-hypotheses $H_{i_1 i_2 \ldots i_k}$, given by

$$P(d|H_{i_1 i_2 \ldots i_k}, {\rm I}) = \int d\boldsymbol{\theta}\ d\delta\chi_{i_1} \ldots d\delta\chi_{i_k}\ P(d|\boldsymbol{\theta}, \delta\chi_{i_1}, \ldots, \delta\chi_{i_k}, H_{i_1 i_2 \ldots i_k}, {\rm I})$$
$$\times\ P(\boldsymbol{\theta}|H_{i_1 i_2 \ldots i_k}, {\rm I}) P(\delta\chi_{i_1}, \ldots, \delta\chi_{i_k}|H_{i_1 i_2 \ldots i_k}, {\rm I}), \tag{7.37}$$

where $\boldsymbol{\theta}$ denotes the regular GW parameters, and we assume that the prior on the GW parameters $\boldsymbol{\theta}$ is independent of the prior on deviations $\{\delta\chi_{i_1} \ldots \delta\chi_{i_k}\}$. Furthermore, we need to calculate the evidence for the hypothesis $\mathcal{H}_{\rm GR}$, given by

$$P(d|\mathcal{H}_{\rm GR}, {\rm I}) = \int d\boldsymbol{\theta}\; P(d|\boldsymbol{\theta}, \mathcal{H}_{\rm GR}, {\rm I}) P(\boldsymbol{\theta}|\mathcal{H}_{\rm GR}, {\rm I}). \tag{7.38}$$

These integrals can be conveniently evaluated by using the Nested Sampling algorithm, described in Sect. 5.4. More specifically, an implementation tailored to ground-based observations of CBC systems by Veitch and Vecchio [7–9] in the LAL [10] was used. The Nested Sampling algorithm gives us the Bayes factors for the various signal hypotheses (either $H_{i_1 i_2 \ldots i_k}$ or $\mathcal{H}_{\rm GR}$) against the noise-only hypothesis $\mathcal{H}_N$ (*cf.* Sect. 5.4.4), given by

$$\begin{aligned} B_N^{i_1 i_2 \ldots i_k} &= \frac{P(d|H_{i_1 i_2 \ldots i_k}, {\rm I})}{P(d|\mathcal{H}_N, {\rm I})}, \\ B_N^{\rm GR} &= \frac{P(d|\mathcal{H}_{\rm GR}, {\rm I})}{P(d|\mathcal{H}_N, {\rm I})}. \end{aligned} \tag{7.39}$$

These can be combined to give the Bayes factors of Eq. (7.16) through

$$B_{\rm GR}^{i_1 i_2 \ldots i_k} = \frac{B_N^{i_1 i_2 \ldots i_k}}{B_N^{\rm GR}}. \tag{7.40}$$

At this point it is worth commenting on the mutual relationships of the hypotheses $H_{i_1 i_2 \ldots i_k}$ among each other, and with $\mathcal{H}_{\rm GR}$. As an example, let us discuss the relationship between H_1 and $\mathcal{H}_{\rm GR}$. Consider the numerator of the Bayes factor $B_{\rm GR}^1$, given by

$$P(d|H_1, {\rm I}) = \int d\boldsymbol{\theta}\; d\chi_1 P(d|\boldsymbol{\theta}, \delta\chi_1, H_1, {\rm I}) P(\boldsymbol{\theta}|H_1, {\rm I}) P(\delta\chi_1|H_1, {\rm I}). \tag{7.41}$$

The parameter space $\{\boldsymbol{\theta}\}$ of the GR waveforms has a natural embedding into the parameter space $\{\boldsymbol{\theta}, \delta\chi_1\}$ of the waveforms associated to H_1, i.e. it can be identified with the hypersurface $\delta\chi_1 = 0$. We could have explicitly excluded this hypersurface from $\{\boldsymbol{\theta}, \delta\chi_1\}$ by setting a prior on $\delta\chi_1$ of the form $P'(\delta\chi_1|H_1, {\rm I}) = 0$ if $\delta\chi_1 = 0$ and $P'(\delta\chi_1|H_1, {\rm I}) =$ const otherwise. However, this would not have made a difference in the integral above. Indeed, with respect to the integration measure induced by the prior probability density on $\{\boldsymbol{\theta}, \delta\chi_1\}$, the surface $\delta\chi_1 = 0$ constitutes a set of measure zero.

Now look at the denominator in the expression for $B_{\rm GR}^1$, which is the evidence for the GR hypothesis, given in Eq. (7.38). Despite the fact that the GR waveforms form a set of measure zero within the set of waveforms associated to the sub-hypothesis H_1, the above integral is clearly not zero. It is the evidence for a qualitatively different hypothesis, whose parameter space $\{\boldsymbol{\theta}\}$ carries a different integration measure with respect to which the marginalisation of the likelihood is carried out. Thus, a sharp distinction should be made between probability densities on *parameter spaces*, and probabilities related to *hypotheses*. Similarly, although the parameter spaces associated with sub-hypotheses $H_{i_1 i_2 \ldots i_k}$ and $H_{j_1 j_2 \ldots j_l}$ may have parameters in common, in each of these spaces the common hypersurface has measure zero unless $H_{i_1 i_2 \ldots i_k} = H_{j_1 j_2 \ldots j_l}$, hence it has no bearing on the calculation of evidences and Bayes factors.

References

1. C. Van Den Broeck, A.S. Sengupta, Binary black hole spectroscopy. Class. Quantum Gravity **24**(5), 1089 (2007)
2. C. Van Den Broeck, A.S. Sengupta, Phenomenology of amplitude-corrected post-Newtonian gravitational waveforms for compact binary inspiral: I. Signal-to-noise ratios. Class. Quantum Gravity **24**(1), 155 (2007)
3. A. Buonanno et al., Comparison of post-Newtonian templates for compact binary inspiral signals in gravitational-wave detectors. Phys. Rev. D **80**, 084043 (2009)
4. R. O'Shaughnessy, C. Kim, Pulsar binary birthrates with spin-opening angle correlations. Astrophys. J. **715**, 230–241 (2010)
5. D.R. Lorimer, Binary and Millisecond Pulsars. Living Reviews in Relativity 8.7 (2005)
6. J. Abadie et al., Calibration of the LIGO gravitational wave detectors in the fifth science run. Nucl. Instrum. Methods Phys. Res. A **624**, 223–240 (2010)
7. J. Veitch, A. Vecchio, Bayesian approach to the follow-up of candidate gravitational wave signals. Phys. Rev. D **78**, 022001 (2008)
8. J. Veitch, A. Vecchio, Assigning confidence to inspiral gravitational wave candidates with Bayesian model selection. Class. Quantum Gravity **25**(18), 184010 (2008)
9. J. Veitch, A. Vecchio, Bayesian coherent analysis of in-spiral gravitational wave signals with a detector network. Phys. Rev. D **81**, 062003 (2010)
10. https://www.lsc-group.phys.uwm.edu/daswg/projects/lalsuite.html

Chapter 8
Results

Now that we have defined the details of TIGER, we are in a position to express the goals in Sect. 6.5 more concretely. The first goal is the ability to test the validity of GR in a generic fashion. If we look at the definition of the hypotheses $\mathcal{H}_{\rm GR}$ and $\mathcal{H}_{\rm modGR}$, given on p. 106 and p. 107 respectively, we can deduce that our hypotheses test whether GR is correct or not. This therefore constitutes a generic test of the validity of GR. Simulations will have to show the applicability of TIGER.

The second goal is to cast the problem in the form of Bayesian hypothesis testing. In Sect. 7.1, we showed that through the appropriate definition of the hypotheses, the problem can be expressed as a hypothesis testing problem. On the practical side, simulations will have to show that the odds ratio, given in Eq. (7.21) for individual sources and Eq. (7.26) for catalogues of sources, is an informative quantity in our search for departures from GR. In Sect. 8.2 we show that the odds ratio is indeed a powerful discriminator between GR being the correct theory or not.

The third goal is to construct a method that is not only generic in its ability to detect possible deviations from GR, but also generic in its use of waveforms and accuracy. Although we have used the phase as an example in the derivation of TIGER, it should be stressed that deviations in any parameterizable part of the waveform can be considered (e.g. amplitude, energy or luminosity). TIGER is also generic in the sense that the 'testing coefficients' are left to the choice of the user. This choice will be influenced by the kind of precision one seeks, as well as the computational cost. Parameters which can not be measured in the specific setting (e.g. advanced detectors) can be left out of the test, without destroying the applicability of TIGER. The choice of the number of testing coefficients will be discussed in Sect. 8.4.

In dealing with data from the advanced detectors, we expect to mainly see sources with a low SNR. The fourth goal of TIGER is to avoid a framework which requires excessive use of free parameters. Not only does the inclusion of extra free parameters degrade the measurability of other (correlated) parameters, it will run the risk of being disfavoured due to Occam's razor, described in Sect. 4.3.2. By splitting $\mathcal{H}_{\rm modGR}$ into the sub-hypotheses $H_{i_1 i_2 \ldots i_k}$, we tackle this problem whilst remaining as general as possible. Most sub-hypotheses only pertain to a small subset of the set of testing coefficients, making them more sensitive at low SNR. Yet, we combine all the sub-

T.G.F. Li, *Extracting Physics from Gravitational Waves*,
Springer Theses, DOI 10.1007/978-3-319-19273-4_8

hypotheses into a statement which is as general as the number of sub-hypotheses considered. The pivotal thing to show in the simulations is that the odds constructed in Eqs. (7.21) and (7.26) has a better performance than the most inclusive sub-hypothesis, i.e. having all phase coefficients simultaneously free. In other words, the disjunction of the propositions "the phase coefficient ψ_i is not as predicted by GR" should perform better than the more restricted conjunction of these propositions. In Sect. 8.2, examples are shown that this is indeed the case.

Finally, the last goal is to have a framework that allows one to coherently analyse information from different sources. Sect. 8.3 shows that increasing the number of sources in a catalogue does indeed increase our confidence in a deviation, if there is one to be found.

8.1 Simulation Details

In Sect. 7.3, we have already seen that the odds ratio in Eq. (7.21) or Eq. (7.26) can be obtained by using the Nested Sampling algorithm (see Sect. 5.4) to calculate the evidence integrals in Eqs. (7.37) and (7.38). The form of the likelihood is specified in Eq. (5.67). To construct the data s, we must simulate the noise n and the signal waveform h_S [*cf.* Eq. (5.65)]. Furthermore, we must specify the prior on the model parameters associated to $\mathcal{H}_{\rm GR}$ and $H_{i_1 i_2 \ldots i_k}$, given by $P(\boldsymbol{\theta}|\mathcal{H}_{\rm GR}, \mathrm{I})$ and $P(\boldsymbol{\theta}, \delta\chi_{i_1}, \ldots, \delta\chi_{i_k}|H_{i_1 i_2 \ldots i_k}, \mathrm{I})$ respectively. Finally, in order to calculate the relevant statistics, i.e. the FAP in Eq. (7.27) and the efficiency in Eq. (7.32), we must specify the simulation population κ for GR signals and the simulation population κ' for a specific alternative theory $\mathcal{H}_{\rm alt}$.

8.1.1 Source Distribution

Firstly, we specify the source distribution for the signal parameters $\boldsymbol{\theta}_{\rm S}$, associated to both κ and κ'. As suggested by Fisher matrix calculations, such as those of Mishra et al. [1], methods based on measuring phase coefficients will be the most accurate at low total mass. That is why we limit ourselves to BNS sources, for which spin will be negligible [2], as well as sub-dominant signal harmonics [3, 4]. Since we will assume a network of advanced detectors, the merger and ringdown signals (see Sect. 3.3.1) are not in the sensitive region of the detector and therefore will not have a large impact on our inference [5]. Consequently, we only consider the inspiral part of the waveform.

The BNS systems are distributed uniformly in volume, with random sky positions and orientations, and their total number is taken to be on the conservative side of the 'realistic' estimates for the number of detectable sources in a 1-year time span [6]. We take the individual NS masses to lie uniformly between 1 and 2 $M_\odot$, in accordance with the range of the NS mass distribution [7]. The distance interval

Table 8.1 Summary of the source distribution used to obtain the results in Sects. 8.2–8.4

Variable	Form	Range	Motivation
Mass 1, m_1	Const	$[1, 2]\,M_\odot$	Neutron star
Mass 2, m_2	Const	$[1, 2]\,M_\odot$	Neutron star
Distance, D_l	$\propto D_l^2$	[100, 400] Mpc	Uniform in volume[a]
Orbital inclination, ι	$\propto \cos\iota$	$[0, \pi]$	Uniform in orientation
Polarisation, ψ	Const	$[0, 2\pi]$	Uniform in orientation
Latitude, θ	$\propto \cos\theta$	$[0, \pi]$	Uniform in sky
Longitude, ϕ	Const	$[0, 2\pi]$	Uniform in sky
Coalescence phase, ϕ_c	Const	$[0, 2\pi]$	Uniform in phase

[a] $D_l = 1000$ Mpc corresponds to $z = 0.21$. For such redshifts, we approximate space as Euclidean and therefore the volume scales as $\propto D_l^2$. For greater redshifts, this relationship breaks down and a prior distribution proportional the co-moving volume ought to be considered instead

is between 100 and 400 Mpc; the former number being the radius within which one would expect ~0.5 BNS signals per year, and 400 Mpc being the approximate maximum distance that Advanced LIGO/Virgo can probe, given a network SNR threshold of 8, which corresponds to the detection threshold set by the LIGO-Virgo collaboration [8]. The corresponding signals are added coherently to simulated noise (assumed to be stationary and modelled by a Gaussian stochastic process) for the Advanced Virgo interferometer and the two Advanced LIGOs (see Fig. 3.4 for their PSDs). A summary of the source distribution is shown in Table 8.1.

Furthermore, we impose a lower cut-off of $\rho_{\text{net}} = 8$ on the network SNR given in Eq. (3.27). This threshold is consistent with the threshold to claim a detection of a GW, set by the LIGO/Virgo collaboration [9]. The analysis of the surviving signals is performed with an appropriately modified version of the Nested Sampling code available in LAL, as described in Sect. 7.3.

Finally, we take the signal waveform to be `TaylorF2` with the phase up to only 2PN order. Despite its simplicity, `TaylorF2` is known to be in good agreement with waveforms that in turn closely match numerical simulations (*cf.* Figs. 3 and 4 in Buonanno et al. [10]). Future development might include more accurate waveforms in order to perform a more accurate test of GR. However, due to its good accordance with numerical relativity for the inspiral part of the waveform, the results shown here can be viewed as a good indication of the sensitivity to deviations from GR in the advanced-detector era.

8.1.2 Method Parameters

Similar to the signal waveform, our model waveforms are also chosen to be in the `TaylorF2` family with the phase up to only 2PN order. As an example, we consider the case of $N_T = 3$ testing coefficients, given by ψ_1, ψ_2 and ψ_3. TIGER thus requires the calculation of $2^3 - 1 = 7$ Bayes factors, i.e. B^1_{GR}, B^2_{GR}, B^3_{GR}, B^{12}_{GR}, B^{13}_{GR}, B^{23}_{GR}, and B^{123}_{GR}. These can be combined to form the odds ratio given in Eq. (7.21) or Eq. (7.26).

Next, we need to assign the priors on the model parameters. For the $\mathcal{H}_{\mathrm{GR}}$ hypothesis, these are the GW parameters $\boldsymbol{\theta}$, *cf.* Eqs. (2.72) and (3.1). The $\mathcal{H}_{\mathrm{modGR}}$ hypothesis (composition of the sub-hypotheses) has, besides the source parameters $\boldsymbol{\theta}$, also the parameters associated to the permitted deviations, given by $\delta\chi_i$.

For the distributions of the GW parameters, we use the same functional forms and limits as given in Veitch and Vecchio [11], except for the distance being allowed to vary between 1 and 1000 Mpc. Specifically, for the sky location and the orientation of the orbital plane we choose uniform priors on the corresponding unit spheres. For the phase at coalescence ϕ_c we choose a flat prior with $\phi_c \in [0, 2\pi]$, and the time of coalescence t_c is restricted to a time interval of 100 ms, accommodating for the largest difference between the time of arrival of the GW in different detectors. The prior on η is flat on the interval [0, 0.25]. For chirp mass we use an approximation to the Jeffreys prior which gives $p(\mathcal{M}|\mathrm{I}) \propto \mathcal{M}^{-11/6}$ (see Veitch and Vecchio [11] for motivation). In addition, component masses are restricted to the interval $m_1, m_2 \in [1, 34]\, M_\odot$, which is the mass range which stellar mass BHs are believed to be in [12]. Especially for BNS, which are the sources we consider in our simulations, this range of the component masses is sufficient. Also, this interval is in accordance with the definition of 'low mass' in standard CBC search pipelines (see e.g. [13]). For the deviations $\delta\chi_i$, we take the priors to be flat and centred on zero, with a total width of 0.5. This will be much larger than the deviations we will use for the simulated signals and hence suffices to illustrate TIGER. For real measurements, one may want to choose a still wider prior in order to achieve a greater level of generality. The choice of priors is summarised in Table 8.2.

Table 8.2 Summary of the priors for the model waveforms used in all the simulations shown in this chapter

Variable	Form	Range	Motivation
Chirp mass, $\mathcal{M}$	$\propto \mathcal{M}^{-11/6}$	$[0.87, 29.60]$ [a]	Jeffreys prior
Symmetric mass ratio, η	Const	$[0, 0.25]$	Indifference
Distance, D_l	$\propto D_l^2$	$[1, 1000]$ Mpc	Uniform in volume[b]
Orbital inclination, ι	$\cos\iota$	$[0, \pi]$	Uniform in orientation
Polarisation, ψ	Const	$[0, 2\pi]$	Uniform in orientation
Latitude, θ	$\cos\theta$	$[0, \pi]$	Uniform in sky
Longitude, ϕ	Const	$[0, 2\pi]$	Uniform in sky
Coalescence phase, ϕ_c	Const	$[0, 2\pi]$	Indifference
Coalescence time, t_c	Const	$[-50, 50]$ ms[c]	Coincidence window
Deviation, $\delta\chi_i$	Const	$[-0.25, 0.25]$	Indifference

[a]Based on component masses in the range $m_1, m_2 \in [1, 34]$
[b]See footnote on p. 123
[c]Assuming that the coalescence time is chosen to be $t_c = 0$

As shown in Sects. 7.2.1 and 7.2.2, both the FAP and the efficiency are invariant under the choice of α. Therefore, for the purpose of presenting the results, the factor α in Eqs. (7.21) and (7.26) will be set to unity.

8.1.3 Example Deviations

Finally, in order to gauge TIGER's performance, we must specify some particular deviation from GR, denoted by the hypothesis $\mathcal{H}_{\text{alt}}$. The deviations considered in this work can be divided into two categories. The first consists of deviations in one of the phase coefficients characterising the `TaylorF2` waveform. The deviations were inserted in a similar fashion as the deviations allowed by our sub-hypotheses, i.e. Eqs. (7.35) and (7.36). Examples will be shown with shifts in ψ_3 and ψ_4. These correspond to the first order in which the dynamical non-linearities of GR are visible [14, 15], and the PN order at which the corrections to the phase due to a modified Einstein-Hilbert action containing terms that are quadratic in the Riemann tensor are visible [16–18], respectively. The deviation in ψ_3 also corresponds to a basic test in which the deviation is in one of the testing coefficients. This example will give us the idealised case in which the deviation can be accommodated for by some of our model waveforms. The deviation in ψ_4 is an example in which the deviation is in a PN order, but one that cannot be accounted for by our model waveforms. We show that deviations in the phase coefficients, whether they are included in the testing coefficients or not, can be detected by TIGER, provided the magnitude of the deviation is sufficiently large.

The second category contains two examples of deviations that cannot be attributed to a deviation in any of the PN coefficients. In the first example, we insert a deviation between successive PN orders. The second example is one in which the deviation not only has a mass dependent coefficient (as with the coefficients of the PN expansion), but the deviation also has a frequency dependence that is a function of the total mass. This category exemplifies the potentially very complex way in which deviations from GR may occur and serves to illustrate the true generality of TIGER. We show that the effects of both of these general examples can be detected, provided that the phase shift induced is large enough (later on we will quantify what large means in this context).

A summary of the deviations considered in this work can be seen in Table 8.3. This table shows the deviations, the form of the deviations, corresponding coefficients and the relevant subsection in which these examples are considered.

Table 8.3 Summary of the deviations considered, their forms, corresponding coefficients and the relevant subsections in which these deviations are considered

Class	Functional form	Coefficient	Section
PN term	Eq. (7.35)	$\delta\chi_3 = 0.1$	8.2.2
		$\delta\chi_3 = 0.025$	8.2.3, 8.3, 8.4
		$\delta\chi_4 = 0.2$	8.2.4
Non PN term	$\frac{3}{128\eta}\delta\chi_A(\pi M f)^{-5/6}$	$\delta\chi_A = -2.2$	8.2.5
	$\frac{3}{128\eta}\delta\chi_{A2}(\pi M f)^{-2+\frac{M}{(3M_\odot)}}$	$\delta\chi_{A2} = 1$	8.2.6

8.2 Odds Ratios and Bayes Factors

8.2.1 Constructing the Background

Before a specific deviation can be considered, it is necessary to construct the background distribution by analysing the odds ratios for a collection of sources, κ, that correspond to the GR hypothesis, $\mathcal{H}_{\mathrm{GR}}$. Sources were distributed according to Table 8.1 and their signals were embedded in the simulated Gaussian and stationary noise, according to the Advanced LIGO/Virgo noise curves, as shown in Fig. 3.4 . The odds ratios for both individual sources as well as catalogues of 15 sources are shown in Fig. 8.1.

For individual sources, as shown in Fig. 8.1a, the distribution has a peak value at $\ln O_{\mathrm{GR}}^{\mathrm{modGR}} < 0$. This means that for GR signals, the GR hypothesis is indeed favoured. We also see that the distribution does extend beyond $\ln O_{\mathrm{GR}}^{\mathrm{modGR}} = 0$, showing that noise can indeed push the odds ratio to favour $\mathcal{H}_{\mathrm{modGR}}$. Therefore, a proper interpretation of the measured odds ratio will always have to take the background into account. The distribution of the log odds ratio for the catalogues of 15 sources, shown in Fig. 8.1b, has the same characteristics, albeit that there is more confidence towards $\mathcal{H}_{\mathrm{GR}}$.

Associated with this background and a maximum FAP of $\beta_{\max} = 0.05$ is a threshold for single sources of $\ln O_\beta = 2$ and a threshold for a catalogue of 15 sources of $\ln \mathcal{O}_\beta = 7$. The measured odds ratio will therefore have to exceed these numbers in order for a detection to be claimed with a FAP of at most 5 %.

Furthermore, this background will be crucial in assessing the efficiencies, defined in Eq. (7.32), for the various example deviations in Sects. 8.2.2–8.2.6. The reader will thus see this background re-appear in most of the sections to come.

(a)

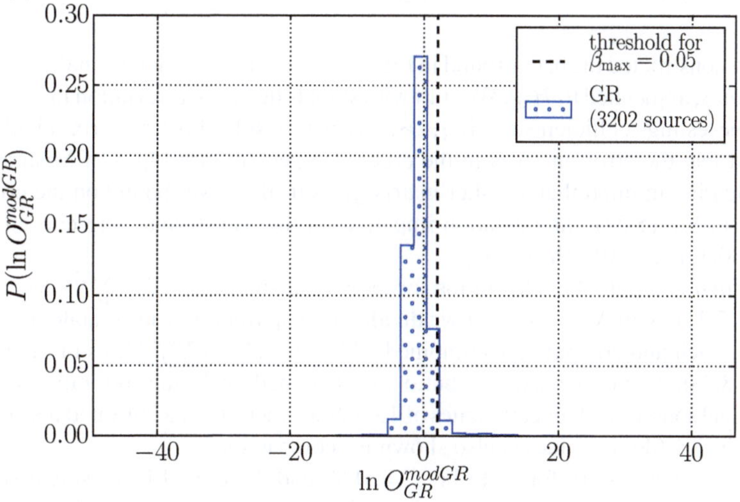

(b)

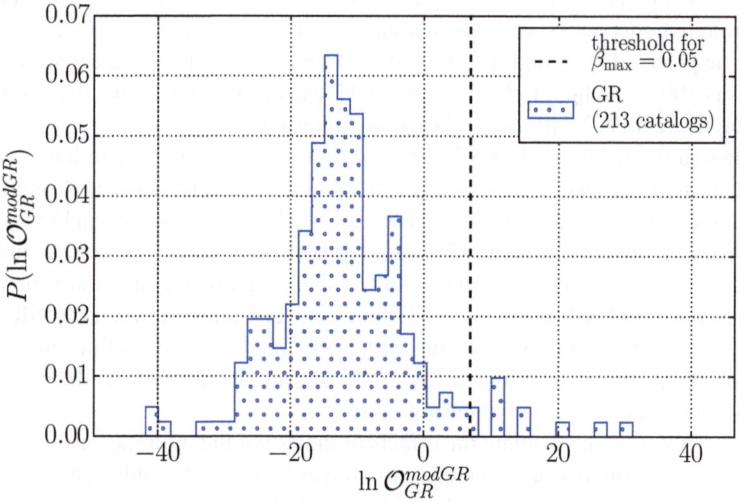

Fig. 8.1 Normalised log odds distributions for a collection of sources associated to the GR hypothesis and distributed according to Table. 8.1. These distributions are the so-called 'background' distributions and will be used to compare the odds ratio obtained from sources which deviate from GR. **a** Log odds ratio distribution for individual sources. For a maximum FAP of $\beta_{max} = 0.05$, the threshold is $\ln O_\beta = 2$. **b** Log odds ratio distribution for catalogues of 15 sources. For a maximum FAP of $\beta_{max} = 0.05$, the threshold is $\ln \mathcal{O}_\beta = 7$

8.2.2 Constant 10 % Shift in ψ_3

Having constructed the background in Fig. 8.1, we are in a position to analyse *specific deviations*, associated to $\mathcal{H}_{\rm alt}$. We start with signals that have a deviation in ψ_3, one of the three testing coefficients, i.e. $\{\psi_1, \psi_2, \psi_3\}$, we consider. Fisher matrix calculations suggest that the error on ψ_3 is in the order of a percent (see Fig. 4 of Mishra et al. [1]). Bearing in mind that a Fisher matrix gives us the lower bound on the error, as shown in Eq. (5.24), the deviation imparted on the waveforms were chosen to be $\delta\chi_3 = 0.1$, i.e. a 10 % shift in ψ_3.

We first compute the odds ratios for individual sources, ${}^{(N_T)}O_{\rm GR}^{\rm modGR}$, according to Eq. (7.21), with $N_T = 3$. Next we divide these up *randomly* into catalogues of 15 sources each and compute the combined odds ratios, ${}^{(N_T)}\mathcal{O}_{\rm GR}^{\rm modGR}$, as in Eq. (7.26). Figure 8.2 shows normalised distributions (red, dashed) of the log odds ratios, both for individual sources and for catalogues of 15 sources each. The background distribution from Fig. 8.1 (dotted, blue) is also shown for comparison.

The efficiencies are found to be $\zeta = 0.58$ and ${}^{(\rm cat)}\zeta = 1.0$ for single sources and catalogues of 15 sources respectively. This means that for a single source, the probability of GR found to be incorrect with a maximum FAP of $\beta_{\rm max} = 0.05$ is about ~60 %. It is evident that combining the odds ratios for sources within a catalogue will strongly boost our confidence in a violation of GR if one is present at the given level. The probability of finding a violation of this magnitude with a catalogue of 15 sources is 100 %. Although the measured FAP can be zero due to the discreteness of the data, the real FAP will always be greater than zero.

Although the results in Fig. 8.2 look promising, it is instructive to have a closer look at the distribution of odds ratios. To aid our understanding, let us look at the (log) odds ratios for individual sources, as a function of SNR. This is shown in Fig. 8.3. The overwhelming majority of signals has an SNR between 8 and 15, which is consistent with our SNR threshold and the placement of sources uniformly in volume up to 400 Mpc. Even for an SNR as low as ~12, there is a separation between the GR signal waveforms and the signal waveforms with a modified ψ_3. This is thus an example that TIGER is indeed sensitive at low SNR. As one would expect, the separation becomes much clearer with increasing SNR.

Next, we want to investigate the effects of the use of the multiple sub-hypotheses $H_{i_1 i_2 \ldots i_k}$. It is useful to look at which of the Bayes factors of the sub-hypotheses tend to give the largest contribution to the odds ratio. In Fig. 8.4, we show the cumulative frequency that a particular $B_{\rm noise}^{i_1 \ldots i_k}$ is the largest, as a function of SNR. The results shown in Fig. 8.4 are entirely as expected, considering that the injected waveform has a shift in ψ_3 only. The Bayes factor $B_{\rm noise}^{3}$, corresponding to the sub-hypothesis H_3, is most often the highest. This signifies that the model H_3 is most likely to be favoured, which is exactly the deviation imposed on the signals. Furthermore, Bayes factors $B_{\rm noise}^{i_1 \ldots i_k}$ corresponding to sub-hypotheses that involve ψ_3 being non-GR, i.e. $B_{\rm noise}^{3}$, $B_{\rm noise}^{13}$, $B_{\rm noise}^{23}$ and $B_{\rm noise}^{123}$, tend to outperform those that do not. Therefore, sub-hypotheses that allow for deviations of the same functional form as the deviation in the signals are favoured over those sub-hypotheses that do not. The model that is

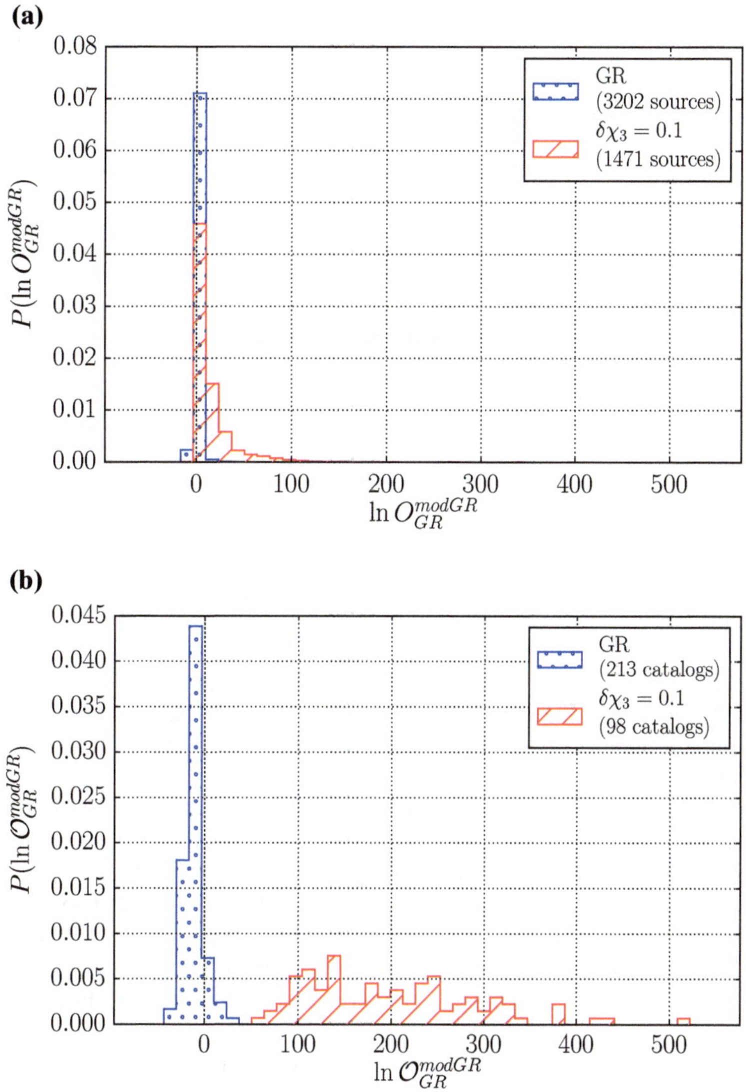

Fig. 8.2 Normalised distributions of log odds ratios for signal with a 10 % shift in ψ_3 (*red, dashed*). This distribution is compared to the background distribution shown in Fig. 8.1 (*blue, dotted*). **a** Log odds ratio distribution for individual sources. The efficiency for individual sources is $\zeta = 0.58$ for a maximum FAP of $\beta_{\max} = 0.05$. **b** Log odds ratio distribution for catalogues of 15 sources. The efficiency in this case is $^{(\text{cat})}\zeta = 1.0$ for $\beta_{\max} = 0.05$. It is evident that the combination of sources greatly improves the ability to find a deviation from GR. A deviation of this form and magnitude can be detected with great confidence if 15 sources are considered

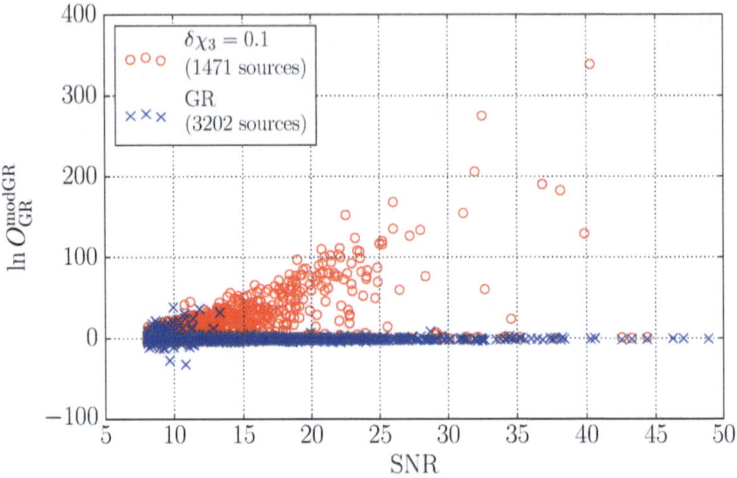

Fig. 8.3 Log odds ratios for individual sources as a function of the SNR. *Blue crosses* represent signals associated with GR waveforms (background), the *red circles* represent signals with a constant 10 % relative offset in ψ_3 (foreground). A separation between the two is visible already for SNR $\gtrsim$ 12 and becomes more pronounced as the SNR increases

ultimately favoured is the model which accounts for the deviations in the signal with the minimum number of parameters, which is consistent with Occam's razor (see Sect. 4.3.2).

Another important point is that most sub-hypotheses perform worse than the GR hypothesis. These sub-hypotheses have extra parameters that do not describe the signal waveform better than the GR hypothesis and are therefore penalised by Occam's razor. For example, if one were to test only with the sub-hypothesis H_2, as a targeted search for a 'massive graviton' would have prescribed, we would have concluded that GR described the signal waveform better. This goes to show that the combination of the sub-hypotheses is important when the nature of the deviation is not known a priori.

We complete the description of this specific deviation from GR by showing two examples of the build-up of Bayes factors and odds ratio within individual catalogues. Fig. 8.5 shows the cumulative Bayes factor for individual sub-hypotheses and the odds ratio as sources with increasingly higher SNR are added to the catalogue.

Both examples show the gradual increase of the Bayes factors as sources are added to the catalogue. This shows that the Bayes factors are not dominated by the most informative source, but instead are built up as sources are added to the catalogue. The first example, shown in Fig. 8.5a, has non-informative sources at low SNR, but as sources with higher SNR are added, the Bayes factors and odds ratio increase. The second example, shown in Fig. 8.5b, shows a similar behaviour except for two things. Firstly, at low SNR, the Bayes factors and odds ratio fall below zero. This is an indication that there are sources for which the GR hypothesis is favoured. This

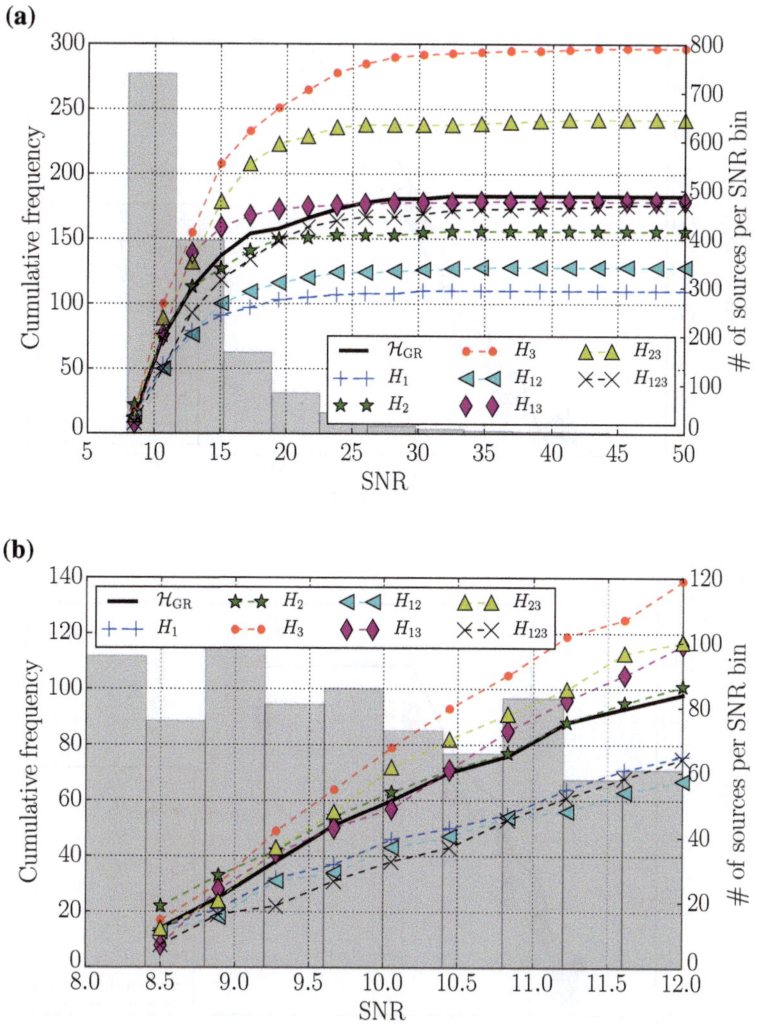

Fig. 8.4 **a** *Curves* and *left vertical axis* For a given SNR, the cumulative frequency that the Bayes factor against noise for a particular sub-hypothesis is the largest, for signals with $\delta\chi_3 = 0.1$. All 1471 simulated sources are included. As expected, $B^3_{\rm noise}$ dominates, and Bayes factors for sub-hypotheses that let ψ_3 be non-GR, tend to outperform those that do not. The GR model outperforms the sub-hypothesis with the largest number of free parameters. *Histogram* and *right vertical axis* Number of sources per SNR bin. **b** The same as above, but restricted to sources with SNR < 12. Similar behaviour as for the full set of 1471 sources is observed. Note that already at SNR close to threshold, the GR hypothesis is more likely to be disfavoured

behaviour is related to the fact that the GR hypothesis is occasionally the dominant hypothesis, as shown in Fig. 8.4, and therefore causes negative log Bayes factors and odds ratio. Secondly, the two highest SNR sources do not contribute significantly to

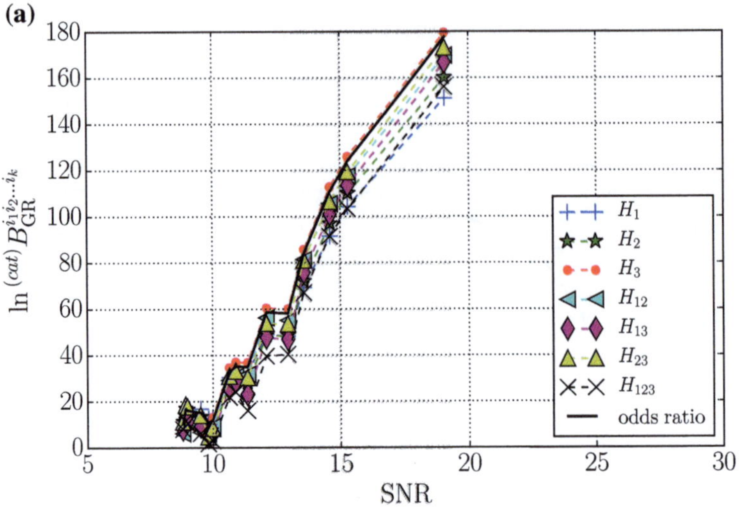

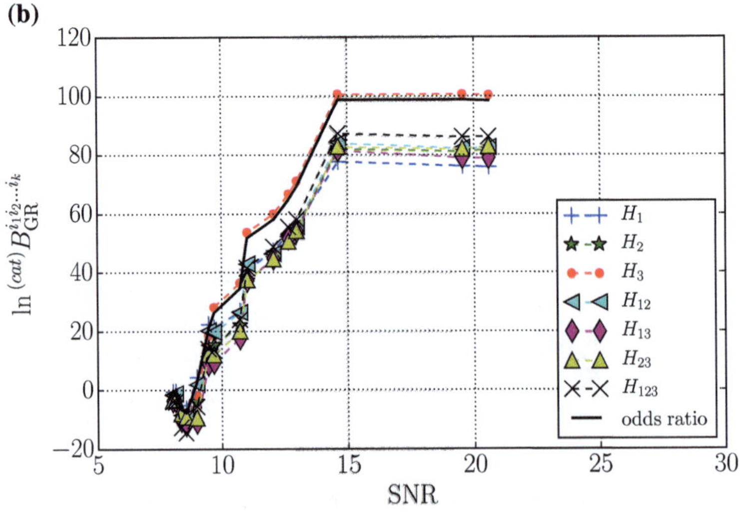

Fig. 8.5 Examples of the cumulative Bayes factors against GR for individual sub-hypotheses within a single catalogue with $\delta\chi_3 = 0.1$. **a** The Bayes factors and the odds ratio gradually grows as the sources with increasingly higher SNR are added. The sub-hypothesis H_3, which matches the nature of the specific deviation from GR, is the most informative sub-hypothesis. **b** Same as panel (**a**), but the Bayes factors and odds ratio initially falls below zero. This happens when the GR hypothesis is favoured, which causes negative log Bayes factors and log odds ratio. Furthermore, the sources with the highest SNRs do not contribute to the odds ratio. Indeed, a high SNR source only gives, *on average*, high Bayes factors and odds ratio

the Bayes factors. This can be understood by considering odds ratios for individual sources as a function of SNR, as shown in Fig. 8.3. As the SNR increases, so does the envelope within which the odds ratio is located. Nevertheless, the SNR is distributed within this envelope due to the presence of noise, and a high SNR only gives, *on average*, a high odds ratio.

Finally, because of the conclusions above, one may be tempted to assign different prior odds to the various sub-hypotheses instead of setting them all equal to each other, as was done in Sect. 7.1. For instance one might consider downweighing the most inclusive sub-hypothesis, H_{123}, by explicitly invoking Occam's razor. However, the violation of GR we assume here is of a rather special form. In reality one will not know beforehand what the nature of the deviation is. In particular, its effect may not be restricted to a single phase coefficient. It is possible that *all* coefficients are affected, in which case one would not want to deprecate H_{123} a priori. In what is to follow, examples will be given that argue that the assignment of different priors to different sub-hypotheses can in fact be detrimental for the outcome of the method.

8.2.3 Constant 2.5 % Shift in ψ_3

It is clear that, if signals arriving at the Advanced LIGO/Virgo network would have a (constant) fractional deviation in ψ_3 as large as 10 %, then, at least under the assumption of Gaussian noise, we would have no trouble in discerning this violation of GR, even if only 15 events were ever recorded. Under such easily detectable deviations, TIGER is shown to behave as expected from Bayesian inference. However, such behaviour might not be present when the deviation is smaller and the violation of GR not completely detectable. It is therefore of interest to consider a smaller deviation in ψ_3; say, 2.5 %.

In Fig. 8.6 we show normalised distributions of the log odds ratio for individual sources, and for catalogues with 15 sources each. For individual sources, the distributions are more or less on top of each other and the efficiency is only $\zeta = 0.11$. The picture is somewhat different for the catalogues. If a catalogue with $\delta\chi_3 = 0.025$ happens to contain one of the sources with a high log odds ratio, visible in Fig. 8.6, then it can boost the combined odds ratio for the catalogue. It therefore comes as no surprise that the efficiency for catalogues of 15 sources is ${}^{(\mathrm{cat})}\zeta = 0.27$.

We now have an example at hand for which the deviation is not as strong as the example considered in Sect. 8.2.2. The interesting question is then whether or not TIGER still works at low SNR, as was the case for $\delta\chi_3 = 0.1$. In Fig. 8.7 we plot the log odds ratios for individual sources against SNR, both for signals with GR waveforms and signals with $\delta\chi_3 = 0.025$. This time the two distributions largely coincide, although there are some sources with a high log odds ratio, which could boost the *combined* odds ratio when they are present in a catalogue of sources. However, for this example, the separation between background and foreground occurs at SNR $\sim$ 15, which is somewhat larger than the $\delta\chi_3 = 0.1$ case, for which the separation comes in at SNR $\sim$ 12. We can conclude, as one expects, that as the deviation

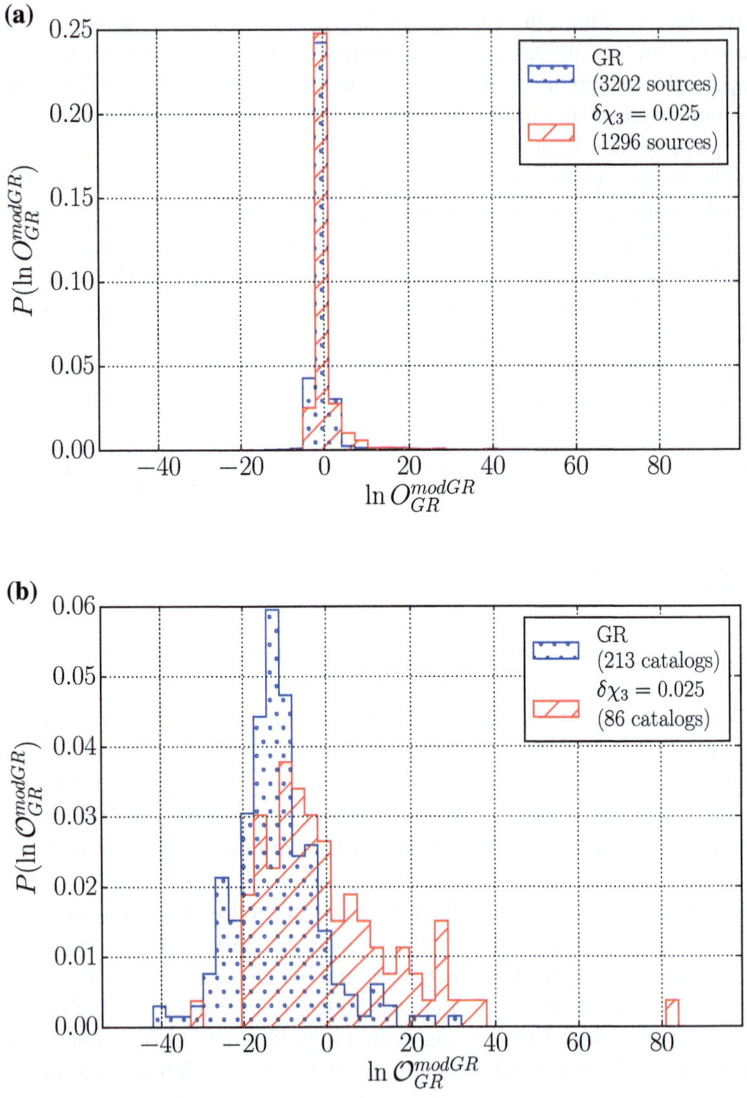

Fig. 8.6 Normalised distributions of odds ratios for sources with $\delta\chi_3 = 0.025$ (*red*, *dashed*) and sources in accordance with GR (*blue*, *dotted*). **a** Normalised distribution of log odds ratios for individual sources. The efficiency in this case is $\zeta = 0.11$, and this particular deviation from GR would be barely detectable. **b** Normalised distribution of log odds ratios for catalogues of 15 sources each. The efficiency is ${}^{(\mathrm{cat})}\zeta = 0.27$, which is an improvement over the efficiency for individual sources. Nevertheless, this specific deviation remains weakly detectable

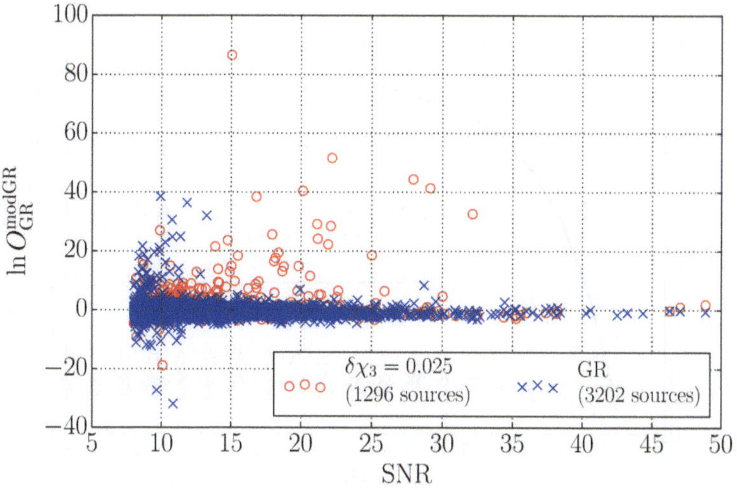

Fig. 8.7 The log odds ratios for individual sources as a function of the SNR. The *blue crosses* represent GR signals, the *red circles* are from signals with a constant 2.5 % relative offset in ψ_3. A separation between the two is marginally visible at SNR $\gtrsim$ 15. Therefore, a deviation of this form and magnitude will be difficult to observe

becomes more difficult to detect, one needs larger SNRs to increase the confidence in a deviation. The more the signal is buried inside the noise, the harder it gets to distinguish small effects.

To gain more insight, we investigate the behaviour for the various sub-hypotheses. Once again, we show, in Fig. 8.8, the cumulative frequency that a particular $B_{\rm noise}^{i_1...i_k}$ is the largest, as a function of SNR, for the case where the signals have $\delta\chi_3 = 0.025$. Unlike for the $\delta\chi_3 = 0.1$ case in Fig. 8.4, $\mathcal{H}_{\rm GR}$ is most often favoured. This means that deviations of GR will rarely be seen if the deviation is of such a magnitude. Furthermore, none of the sub-hypotheses has a better performance than another. If the deviation is weakly detectable, none of the sub-hypotheses perform particularly well, arguing that *all* sub-hypotheses are needed in order to achieve the best sensitivity.

To further illustrate the necessity of combining all sub-hypotheses, we show the build-up of Bayes factors and odds ratio of a representative catalogue with $\ln {}^{(3)}\mathcal{O}_{\rm GR}^{\rm modGR} > 0$. In Fig. 8.9a we show the build-up of the log Bayes factors for the various sub-hypotheses against GR, as well as the odds ratio itself. Several interesting features can be seen in Fig. 8.9a. Firstly, a catalogue consist of informative sources that increase the odds ratio and the Bayes factors, as well as non-informative sources that decrease the odds ratio and the Bayes factors. A high SNR is not predeterminate of an informative source in the case where the deviation is weakly detectable. However, if the source is confidently detectable, we do see from Fig. 8.3 that its confidence increases with SNR. Another thing to note in Fig. 8.9a is the separation of Bayes factors above and below $\ln \mathcal{O}_{GR}^{\rm modGR} = 0$. For example, if one were to test only with the most inclusive sub-hypothesis, i.e. H_{123}, one would have come to the

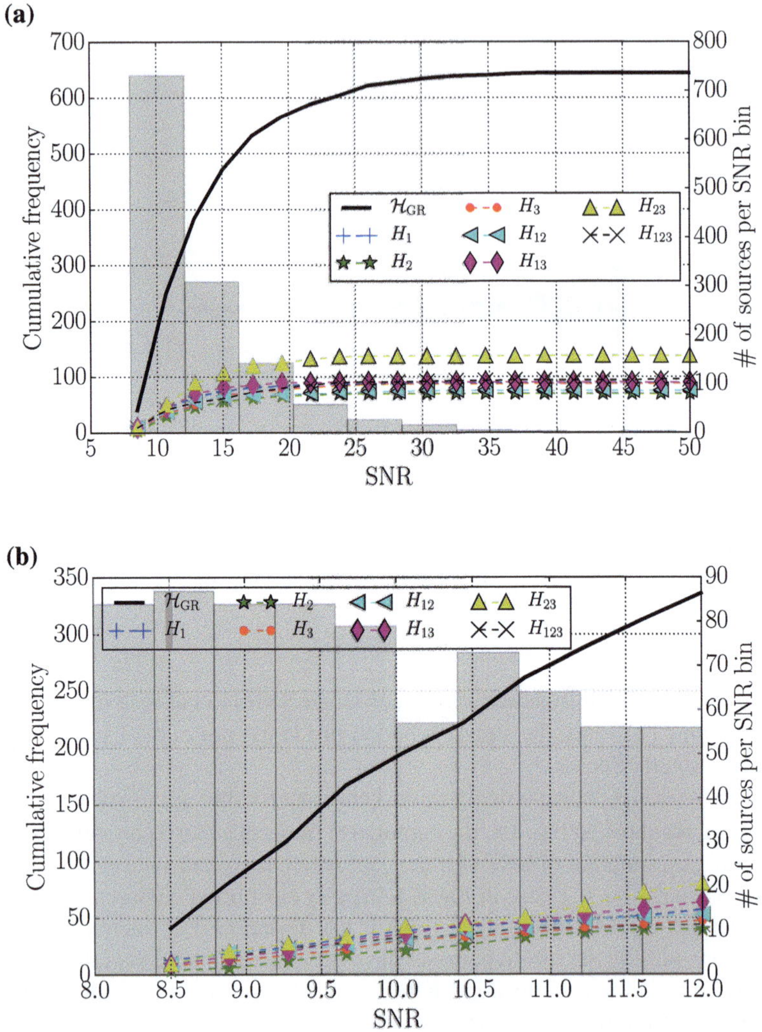

Fig. 8.8 **a** *Curves* and *left vertical axis* For a given SNR, the cumulative frequency that the Bayes factor against noise for a particular sub-hypothesis is the largest, for signals with $\delta\chi_3 = 0.025$. All 1296 simulated sources were used. As opposed to the $\delta\chi_3 = 0.1$ case, $B^{\mathrm{GR}}_{\mathrm{noise}}$ dominates. None of the sub-hypotheses outperforms the other, showing that the inclusion of all sub-hypotheses is crucial in the case where the deviation is only weakly detectable. *Histogram* and *right vertical axis* The number of sources per SNR bin. **b** The same as above, but restricted to sources with SNR < 12. Similar behaviour as for the full set of 1296 sources is observed

conclusion that $\mathcal{H}_{\mathrm{GR}}$ is the favoured model. The same is true if we had only tested H_2, as one would do when specifically looking for a 'massive graviton'. However, considering all sub-hypotheses together, the opposite is true. Another feature worth pointing out is that the sub-hypothesis H_3 is not the most favoured one; instead, it is

H_1. Indeed, not knowing beforehand what the precise nature of the GR violation will be, it can be pivotal to combine all sub-hypotheses instead of considering individual sub-hypotheses.

Finally, in Fig. 8.9b, we also look at a representative catalogue for which the GR hypothesis is favoured, i.e. $\ln \mathcal{O}_{\mathrm{GR}}^{\mathrm{modGR}} < 0$. In this example, all sources have a negative log Bayes factors, except for a single source at SNR $\sim$ 12 that causes a sudden rise. Although such a signature would be a justification for a follow-up investigation into this specific source, the reader is reminded that this single source is inconsistent with the remaining 14 sources in this catalogue. Indeed, the catalogue odds ratio is a statement from *all* sources in the catalogue.

8.2.4 Constant 20 % Shift in ψ_4

In the two previous examples, the deviations were in one of the testing coefficients. The ability to discern a deviation from GR might therefore not surprise the reader, as a significant subset of the sub-hypotheses have a channel through which the deviation can be modelled exactly.

In the next example we aim to investigate a case in which the deviation cannot be modelled exactly by one of the sub-hypotheses. Therefore, we choose the deviation to be in a higher order coefficient than our testing coefficients allow for. In this example, we choose the deviation to be in ψ_4, which is one PN order beyond the highest order that is considered in the testing coefficients. In order to be confident in a detection, we choose the magnitude to be such that the phase shift induced by the deviation at the most sensitive frequency of the Advanced LIGO/Virgo detectors ($f \sim 150$ Hz, see Fig. 3.4) is comparable to the case in which $\delta\chi_3 = 0.1$. For the shift in ψ_4, this amounts to the relative shift being $\delta\chi_4 = 0.2$. Physically, a deviation in ψ_4 can be an indication of models in which the Einstein-Hilbert action includes quadratic curvature terms [16–18].

In Fig. 8.10 we show the odds ratio for individual sources and for random catalogues with 15 sources each. For individual sources, the efficiency is $\zeta = 0.45$, so that the separation between the background and the foreground is present. However, when one assumes random catalogues of 15 sources each, the efficiency increases to ${}^{(\mathrm{cat})}\zeta = 0.97$ and the separation becomes very significant. This further illustrates the importance of combining information from multiple sources.

Figure 8.11 shows the odds ratio as a function of the optimal SNR, both for GR injections and signals with $\delta\chi_4 = 0.2$. Similar to Fig. 8.3 for the $\delta\chi_3 = 0.1$ case, the separation between the background and foreground becomes apparent at SNR $\sim$ 12. However, the odds ratio for the $\delta\chi_4 = 0.2$ case only reaches up to $\ln O_{\mathrm{GR}}^{\mathrm{modGR}} \sim 60$, whereas for $\delta\chi_3 = 0.1$, it reaches up to $\ln O_{\mathrm{GR}}^{\mathrm{modGR}} \sim 350$. Clearly, this can be attributed to the fact that the deviation cannot be modelled exactly by the waveforms associated to the sub-hypotheses. We therefore conclude that although the efficiency is an informative quantity, there is also information in the magnitude of the separation between background and foreground.

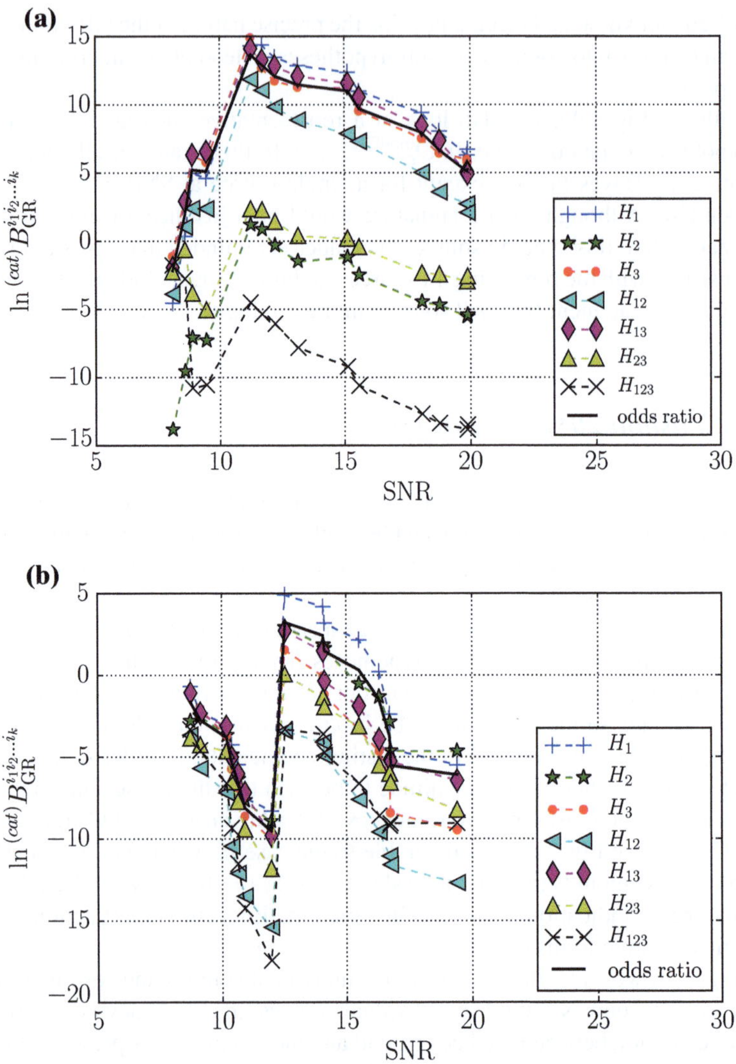

Fig. 8.9 The build-up of cumulative Bayes factors against GR for individual sub-hypotheses, and the odds ratio, for a typical catalogue with $\delta\chi_3 = 0.025$. **a** A catalogue that has $\ln \mathcal{O}_{\mathrm{GR}}^{\mathrm{modGR}} > 0$. Note that on the basis of the Bayes factor against GR of the most inclusive sub-hypothesis, i.e. H_{123}, alone, one would have concluded that the GR model is in fact the favoured one. Even the log Bayes factor for H_{23} ends up being negative. Additionally, the sub-hypothesis with the largest Bayes factor is not H_3 but H_1. This illustrates that it is necessary to include as many sub-hypotheses as possible in the analysis. **b** A representative catalogue with $\ln \mathcal{O}_{\mathrm{GR}}^{\mathrm{modGR}} < 0$. Note the single source at SNR $\sim$ 12 that causes a sudden rise. Although this source can be singled out as a possible signature of a deviation, its result is inconsistent with the remaining sources in the catalogue, causing the catalogue odds ratio to favour the GR hypothesis

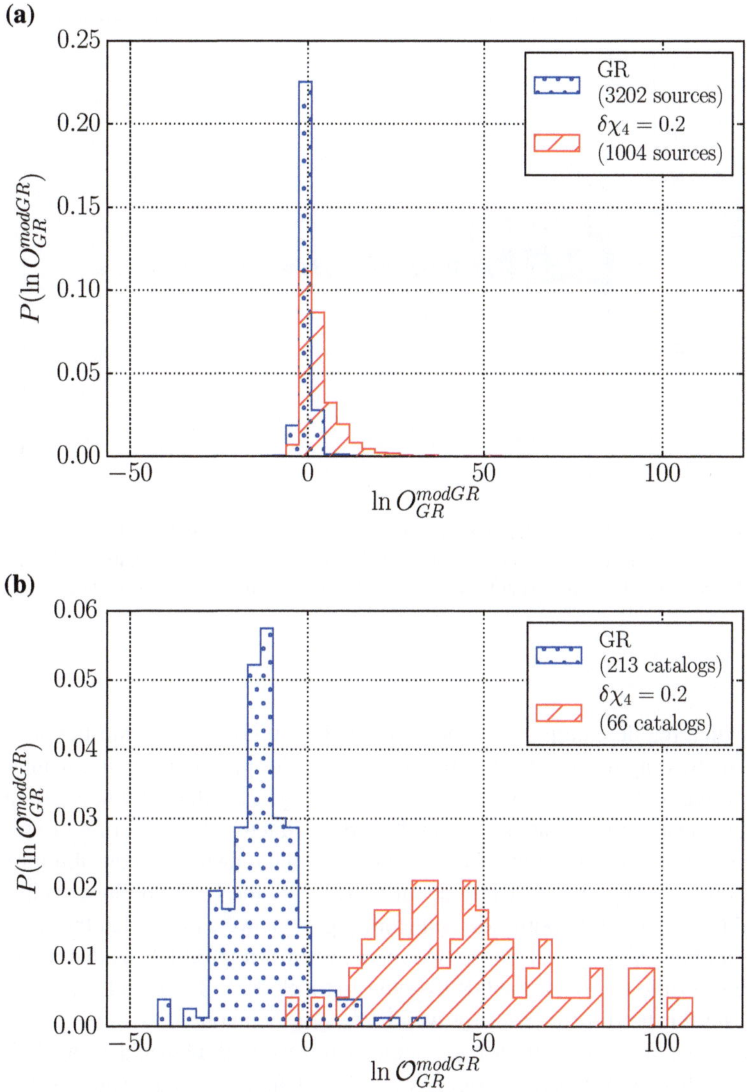

Fig. 8.10 Normalised distributions of odds ratios for sources with $\delta\chi_4 = 0.2$ (*red*, *dashed*) and sources in accordance with GR (*blue*, *dotted*). **a** Normalised distribution of log odds ratios for individual sources. The efficiency in this case is $\zeta = 0.45$, and already shows good performance for individual sources. **b** Normalised distribution of log odds ratios for catalogues of 15 sources each. The efficiency is ${}^{(cat)}\zeta = 0.97$, which shows that such a deviation can be detected with near certainty

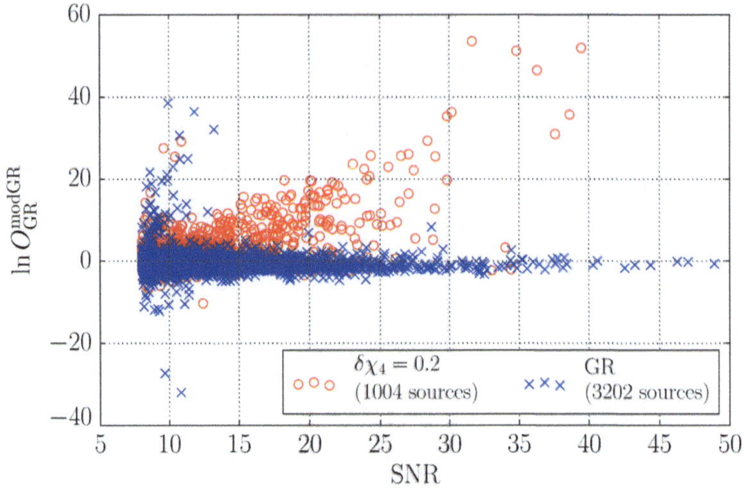

Fig. 8.11 Log odds ratios for individual sources as a function of the SNR. The *blue crosses* represent signals from GR, the *red circles* come from signals with a constant 20 % relative offset in ψ_4. A separation between the two is visible for SNR $\gtrsim$ 12 and becomes more pronounced as the SNR increases

Now that the deviation cannot be accounted for by one of the sub-hypotheses, it will be interesting to see which of the sub-hypotheses provides the dominant contribution to the odds ratio. Therefore, we show, in Fig. 8.12, the cumulative frequency that a given sub-hypothesis has the highest Bayes factor, as a function of the SNR. Looking at Fig. 8.12, we come across a curious behaviour for the contributions from the various sub-hypotheses. Up to about SNR $\sim$ 25, the GR hypothesis is favoured over all the other sub-hypotheses. But from Fig. 8.11, we can see that the separation sets in at about SNR $\sim$ 12. These two pieces of information tell us that although it is more probable to favour the GR hypothesis for SNR $<$ 25, the odds ratio is more significant for the limited cases for which $\mathcal{H}_{\mathrm{modGR}}$ is favoured.

What does a 20 % shift in the ψ_4 mean in terms of physical limits? Current constraints from binary pulsars yield $\delta\chi_4 < 10^7$ [19]. It is evident from Fig. 8.10 that if no deviation is to be found in Advanced LIGO/Virgo, a constraint of $\delta\chi_4 \lesssim 10^{-1}$ can be placed. Therefore, direct measurement of CBC systems can constrain theories of this kind to precisions that have not been seen before.

8.2.5 Non-PN Frequency Contribution; Case One

The aim of Sect. 8.2.2–8.2.4 was to get an impression of the sensitivity of TIGER to deviations in PN coefficients. In order to gauge this, we assumed a constant relative

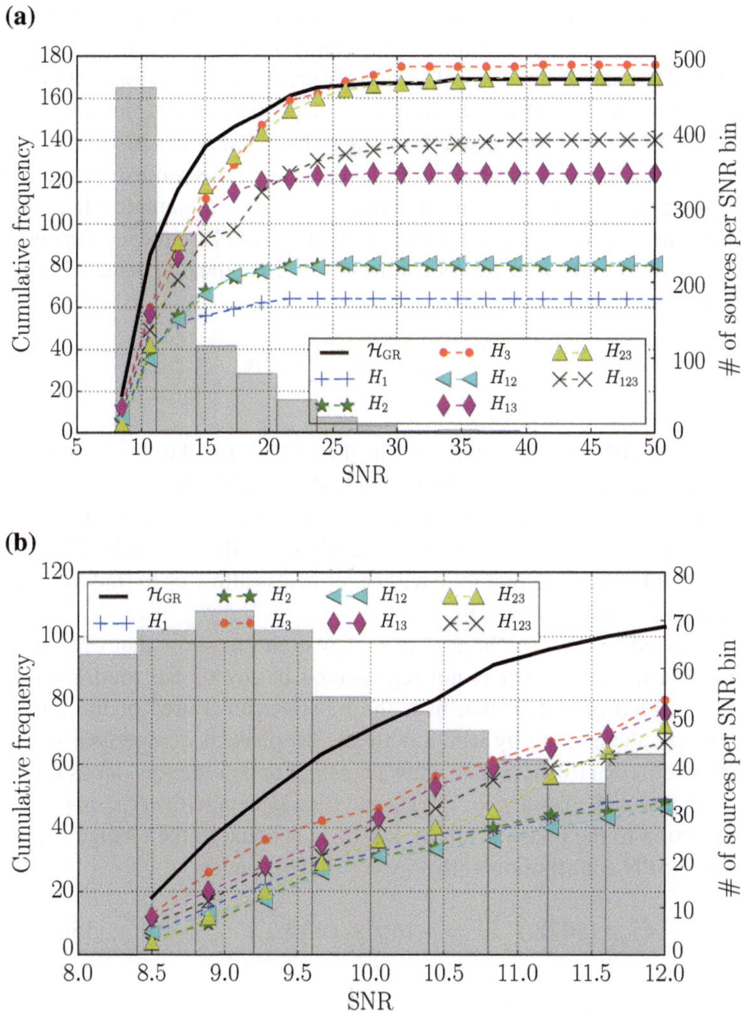

Fig. 8.12 **a** *Curves* and *left vertical axis* For a given SNR, the cumulative frequency that the Bayes factor against noise for a particular sub-hypothesis is the largest, for signals with $\delta\chi_4 = 0.2$. All 1004 simulated sources were used. *Histogram* and *right vertical axis* The number of sources per SNR bin. **b** The same as above, but restricted to sources with SNR < 12. Similar behaviour as for the full set of 1004 sources is observed. In terms of frequency, GR is the favoured model until SNR ~ 20. However, from Fig. 8.11, we see that in terms of the magnitude of the odds ratio, the separation already occurs around SNR ~ 12. It should be stressed that it is the magnitude of the odds ratio that quantifies our (dis)believe in GR, and not the frequency that a hypothesis is preferred

offset in the physically interesting PN coefficients ψ_3 and ψ_4. However, we stress once again that we do not expect a violation of GR to manifest itself as a simple constant relative shift in one of the PN coefficients.

Even if modifications are confined to the PN coefficients, the $\delta\chi_i$ in the signals can be dependent on $(\mathcal{M}, \eta)$, in addition to whatever charges and coupling constants may be present in alternative theories of gravity. Moreover, a deviation from GR could introduce terms in the phase with frequency dependences that do not correspond to any of the PN contributions. We now show that TIGER can also be sensitive to violations of that kind, even though the model waveforms we use in our analyses only have deformations of PN terms. Let us give a heuristic example where the phase of the simulated signals contains a term with an non-PN frequency dependence in between that of the 1PN and 1.5PN contributions. Specifically, the shift is assumed to take the form

$$\frac{3}{128\eta}(\pi M)^{-5/6}\delta\chi_A f^{-5/6}. \tag{8.1}$$

We note that the 1PN term goes like f^{-1} and the 1.5PN term like $f^{-2/3}$, and therefore, the deviation introduced here could be dubbed '1.25PN'. However, for the use of the Nested Sampling algorithm, we will continue to use the same model waveforms as before, which can only have shifts in the phase coefficients at 0.5PN, 1PN, and 1.5PN. Our aim is to show that they will nevertheless allow us to find a deviation in the signal of the form shown in Eq. (8.1).

We now need to make a choice for $\delta\chi_A$. We aim to show that even if there is a deviation in the phase that is not represented in any of our model waveforms, it can be detected provided that, near the most sensitive area of the noise curve ($f \sim 150$ Hz), the amount by which it affects the phase is comparable to a shift in the 1.5PN coefficient by more than a few percent. For definiteness, let us take $\delta\chi_A$ to be constant, and such that at $f = 150$ Hz and for a system with $m_1 = m_2 = 1.5\,M_\odot$, the contribution to the phase given by Eq. (8.1) is equal to the change caused by a shift in the 1.5PN contribution with $\delta\chi_3 = 0.1$. This yields

$$(\pi 3\,M_\odot)^{-5/6}\delta\chi_A(150\,\text{Hz})^{-5/6} = -16\pi \times (\pi 3\,M_\odot)^{-2/3} \times 0.1 \times (150\,\text{Hz})^{-2/3}, \tag{8.2}$$

leading to $\delta\chi_A = -2.2$. The phase shift at $f = 150$ Hz is $\sim$5 radians.

Firstly, we show normalised distributions of the log odds ratios, both for individual sources and for catalogues of 15 sources each in Fig. 8.13. The efficiencies are $\zeta = 0.71$ and ${}^{(\text{cat})}\zeta = 1.0$ for individual sources and for catalogues of 15 sources respectively. As expected, there is an excellent separation for catalogues of sources between the background and the foreground.

As before, we also show the odds ratios of individual sources as a function of SNR in Fig. 8.14. We see that even at small SNR, there is already a good separation between background and the foreground.

In Fig. 8.15, we show the cumulative frequency that the Bayes factor against noise for a particular sub-hypothesis is the largest, arranged with increasing SNR. From SNR $\sim$ 15, the Bayes factor B^{2}_{noise} starts to dominate, followed by B^{23}_{noise} and B^{3}_{noise}, with the latter two crossing over between SNR $\sim$ 20 and SNR $\sim$ 25.

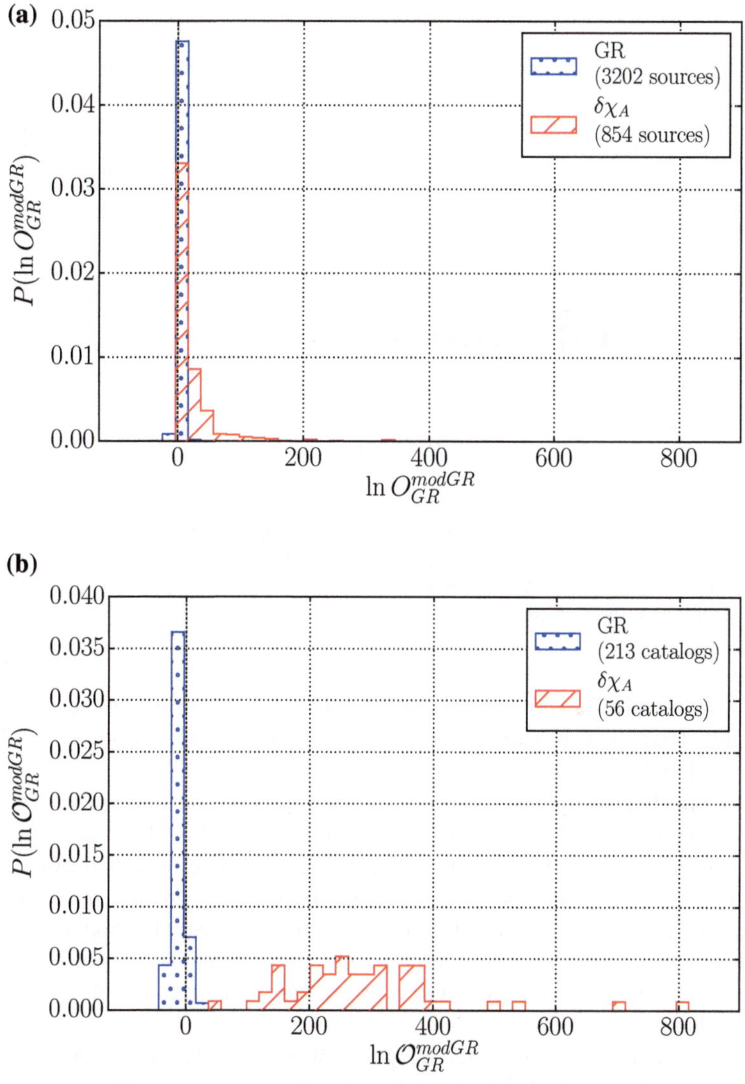

Fig. 8.13 Normalised distributions of odds ratios for sources with a deviation of the form given by Eq. (8.1) and a coefficient $\delta\chi_A = -2.2$ (*red*, *dashed*) and sources in accordance with GR (*blue*, *dotted*). **a** Normalised distribution of log odds ratios for individual sources. The efficiency in this case is $\zeta = 0.71$, and already shows good performance for individual sources. **b** Normalised distribution of log odds ratios for catalogues of 15 sources each. The efficiency is ${}^{(\mathrm{cat})}\zeta = 1.0$. It is evident that TIGER can discern deviations from GR that are not present in one of the PN orders

As the deviation in Eq. (8.1) resides between the PN coefficients ψ_2 and ψ_3, we indeed expect the sub-hypotheses that contain these coefficients to perform the best. Furthermore, already at SNR ~ 9, *all* of the $B_{\mathrm{noise}}^{i_1 i_2 \ldots i_k}$ dominate the Bayes factor $B_{\mathrm{noise}}^{\mathrm{GR}}$

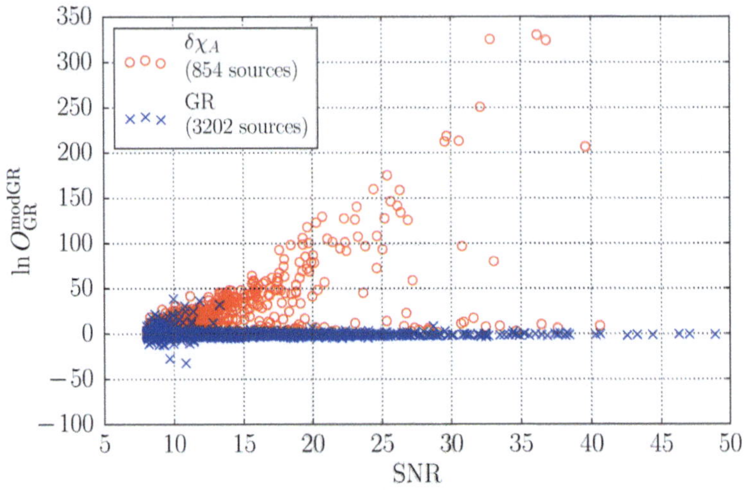

Fig. 8.14 Log odds ratios for individual sources as a function of the SNR. The *blue crosses* represent signals from GR, the *red circles* come from signals with a deviation of the form given by Eq. (8.1) and a coefficient $\delta\chi_A = -2.2$. A separation between the two is visible for SNR $\gtrsim 12$ and becomes more pronounced as the SNR increases

for the GR hypothesis. However, near the SNR threshold, no single sub-hypothesis dominates clearly, which again shows that as many sub-hypotheses as possible should be included in the analysis.

We can further justify the inclusion of all sub-hypotheses by comparing Fig. 8.15 to the similar plot for the $\delta\chi_3 = 0.1$ case, shown in Fig. 8.4. The main difference is that for the $\delta\chi_3 = 0.1$ case, where the deviation can be modelled exactly by a subset of the sub-hypotheses, only a few sub-hypotheses perform better than the GR hypothesis. On the other hand, when the deviation cannot be modelled exactly, all sub-hypothesis perform better than the GR hypothesis. Therefore, if one does not know the alternative theory to GR a priori, it is crucial to include all available sub-hypotheses.

Especially in this case, it is interesting to look at the growth of cumulative Bayes factors against GR for individual sub-hypotheses, as well as of the odds ratio, as sources with increasing SNR are being added within catalogues of 15 sources. This is shown for two example catalogues in Fig. 8.16.

Together with Fig. 8.15, we can conclude that which sub-hypothesis comes out on top will vary from one catalogue to another. In the examples of Fig. 8.16, we see that either H_{12} or H_{23} gives the largest contribution. However, examples can be found where any of the other sub-hypotheses contributes the most. In the catalogue shown in Fig. 8.16b, if one were to favour particular (subsets of) sub-hypotheses a priori, the log odds ratio could be lowered by as much as 100. This could have a large effect on the FAP (*cf.* Fig. 8.13). These are again arguments for using as many sub-hypotheses as possible, and giving them equal relative prior odds.

(a)

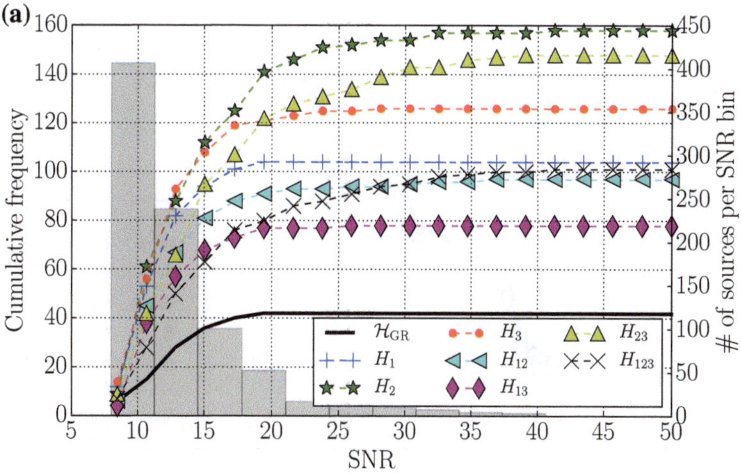

(b)

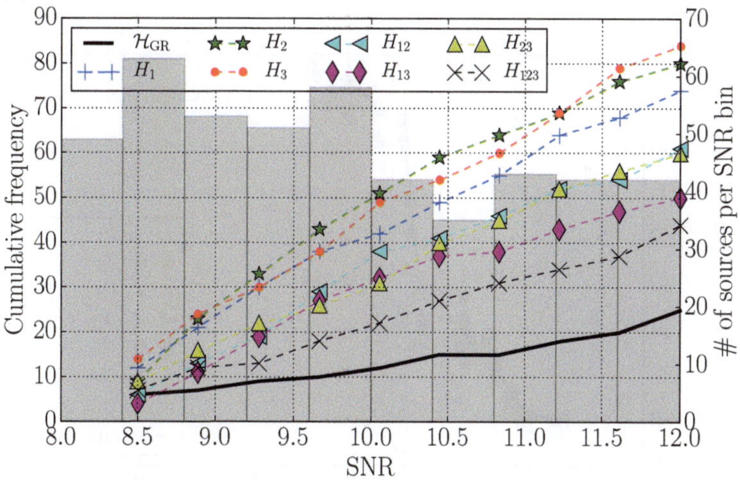

Fig. 8.15 **a** *Curves* and *left vertical axis* For a given SNR, the cumulative frequency that the Bayes factor against noise for a particular sub-hypothesis is the largest, for signals with a deviation of the form given by Eq. (8.1) and a coefficient $\delta\chi_A = -2.2$. All 854 simulated sources were used. The sub-hypotheses allowing for deviations in the PN orders which are the closest to the shift, i.e. 1PN and 1.5PN, are favoured. *Histogram* and *right vertical axis* The number of sources per SNR bin. **b** The same as above, but restricted to sources with SNR < 12. Similar behaviour as for the full set of 854 sources is observed

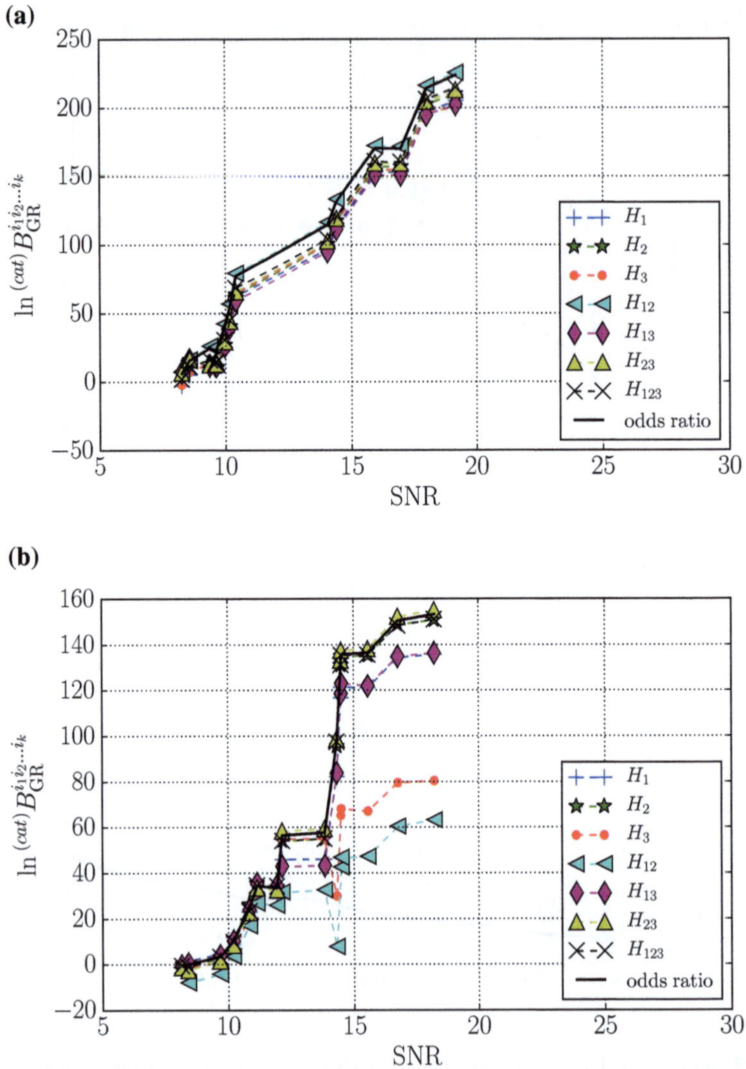

Fig. 8.16 Two examples of how cumulative Bayes factors against GR for individual sub-hypotheses and the odds ratio, grow as sources with increasing SNR are being added within catalogues of 15 sources in total, for a phase shift given in Eq. (8.1) and $\delta\chi_A = -2.2$. Note the large differences in contributions from different sub-hypotheses, and in the ordering of Bayes factors, between these two catalogues. **a** A catalogue where H_{12} is the most informative sub-hypothesis. **b** A catalogue where H_{12} is the least informative sub-hypothesis. In this case, the difference between H_{12} and the most informative sub-hypothesis, H_{23}, is significant

8.2.6 Non-PN Frequency Contribution; Case Two

In our final example, we want to take the generality of the deviation a step further than was done in Sect. 8.2.5. The signals are given, besides the usual mass dependent magnitude [e.g. Eq. (8.1)], a deviation in the phase that has a mass dependent frequency power. Assuming that the deviations are determined only by the component masses, this would constitute the most general case. Specifically, the deviation is of the form

$$\Psi^{\mathrm{GR}}(\mathcal{M}, \eta; f) \rightarrow \Psi^{\mathrm{GR}}(\mathcal{M}, \eta; f) + \frac{3}{128\eta}(\pi M f)^{-2+\frac{M}{3M_{\odot}}}. \tag{8.3}$$

We note that, at $f = 150$ Hz and for a system with component masses in the middle of our range, i.e. $m_1 = m_2 = 1.5\,M_{\odot}$, the change in phase is about the same as for a 10 % shift in ψ_3. Furthermore, in the total mass interval considered in these simulations, the PN order of the deviation ranges effectively from 0.5PN for the least massive systems to 1.5PN for the most massive systems.

In Fig. 8.17, the odds ratios for both individual sources and for catalogues of 15 sources with a deviation as in Eq. (8.3) are again compared to the background distribution from Fig. 8.1. The efficiencies are $\zeta = 0.91$ and ${}^{(\mathrm{cat})}\zeta = 1.0$ for individual sources and catalogues respectively. Given the examples in Sects. 8.2.2 and 8.2.5, it should come as no surprise that the deviation can be found with both single sources and catalogues of 15 sources. We would once again argue that even for the most general mass dependent deviations, TIGER can detect deviations as long as the induced phase shift is comparable to a few percent shift in ψ_3 at the most sensitive region of the advanced detectors ($f \sim 150$ Hz).

Next, we show the distribution of odds ratio as a function of the SNR in Fig. 8.18. The separation, as expected, becomes pronounced close to the threshold value of the SNR.

Furthermore, we can study the behaviour of the Bayes factors for individual sub-hypotheses. Firstly, in Fig. 8.19, we show the cumulative frequency that a particular sub-hypothesis has the highest Bayes factor, as a function of the SNR. For this particular deviation, the most inclusive sub-hypothesis H_{123} is the most informative, but only for SNR $\gtrsim 17$. We can interpret this as a manifestation of Occam's razor. When the signal is weak compared to the noise, the sub-hypothesis H_{123} is penalised by Occam's razor because of it has added complexity compared to the other sub-hypotheses, yet it does not describe the signal significantly better. However, once the signal is strong enough, the sub-hypothesis H_{123} can describe the signal, on average, significantly better than the other sub-hypotheses, and therefore has a higher Bayes factor despite Occam's razor.

Finally, we show two examples of the build-up of Bayes factors for the individual sub-hypotheses and the odds ratio in Fig. 8.20. Both examples show the expected gradual increase in Bayes factors as more sources are added to the catalogue. In Fig. 8.20a, we see an example catalogue where a single source separates the Bayes factors for different sub-hypotheses. This source separates the sub-hypotheses with

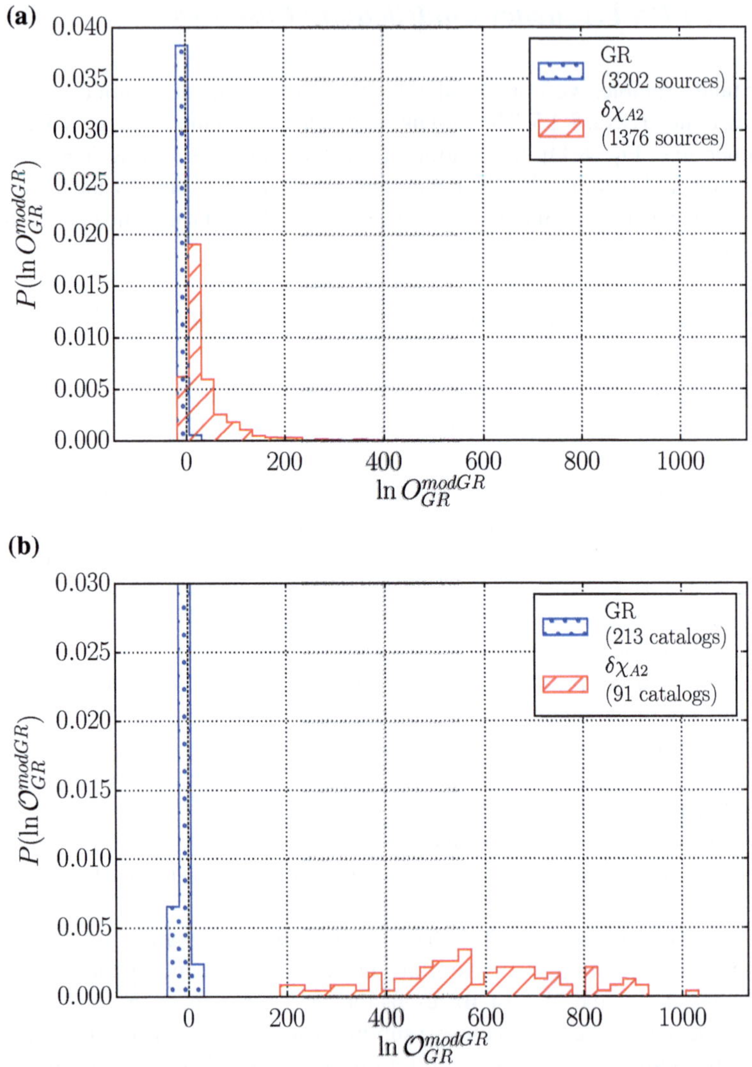

Fig. 8.17 Normalised distributions of odds ratios for sources with a deviation of the form given by Eq. (8.3) (*red, dashed*) and sources in accordance with GR (*blue, dotted*). **a** Normalised distribution of log odds ratios for individual sources. The efficiency in this case is $\zeta = 0.91$, and already shows good performance for individual sources. **b** Normalised distribution of log odds ratios for catalogues of 15 sources each. The efficiency is ${}^{(\mathrm{cat})}\zeta = 1.0$. It is evident that TIGER can even discern deviations that have a mass dependent frequency power

multiple free coefficients from the sub-hypotheses that only have a single free phase coefficient. The complex behaviour of the phase shift in Eq. (8.3) cannot be accurately described by a single free phase coefficient, and the relevant sub-hypotheses

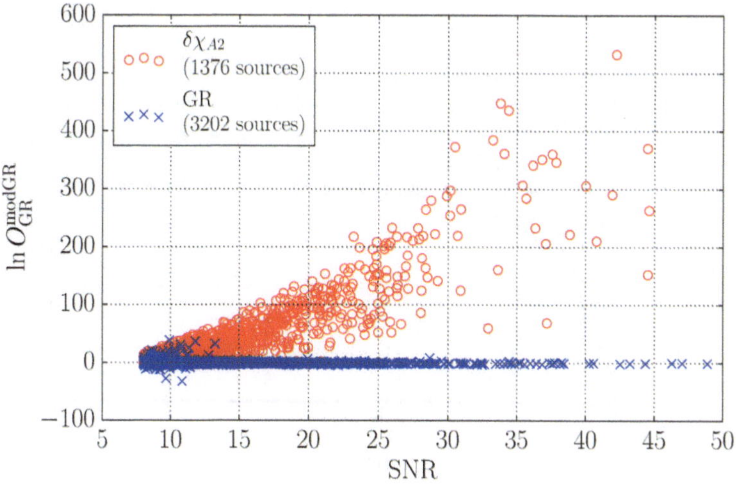

Fig. 8.18 Log odds ratios for individual sources as a function of the SNR. The *blue crosses* represent signals from GR, the *red circles* come from signals with a deviation of the form given by Eq. (8.3). A separation between the two is visible for SNR $\gtrsim$ 12 and becomes more pronounced as the SNR increases

are penalised for this. In Fig. 8.20b, we see an example catalogue where the sub-hypothesis H_{123} start off being the least informative sub-hypothesis. But as sources with increasing SNR are added, the sub-hypothesis H_{123} becomes more favoured compared to the other sub-hypotheses. This is to be compared with what is shown in Fig. 8.19.

8.3 Effect of the Catalogue Size

Consider the example with $\delta\chi_3 = 0.025$, given in Sect. 8.2.3. For a maximum FAP of $\beta_{\rm max} = 0.05$ and 15 sources per catalogue, we have an efficiency of ${}^{(\rm cat)}\zeta = 0.27$. For such an efficiency, we would consider this specific deviation from GR to be hardly detectable. However, the number of sources per catalogue, 15 in the examples shown so far, and the chosen maximum FAP, $\beta_{\rm max} = 0.05$, are somewhat arbitrary. In reality, the size of the catalogue will depend on the number of detected sources, and the FAP is set according to the required confidence. It is therefore of interest to investigate the effects of both of these factors. In Fig. 8.21 we show the behaviour of the efficiency as a function of the catalogue size and the maximum FAP. To account for the arbitrariness in which the sources are combined to form catalogues, we show the median (central curve) and 68 % confidence interval (error bars) of the efficiencies from 5000 different ways in which the sources were ordered into catalogues. Results are shown for $\beta_{\rm max} \in \{0.05, 0.01\}$.

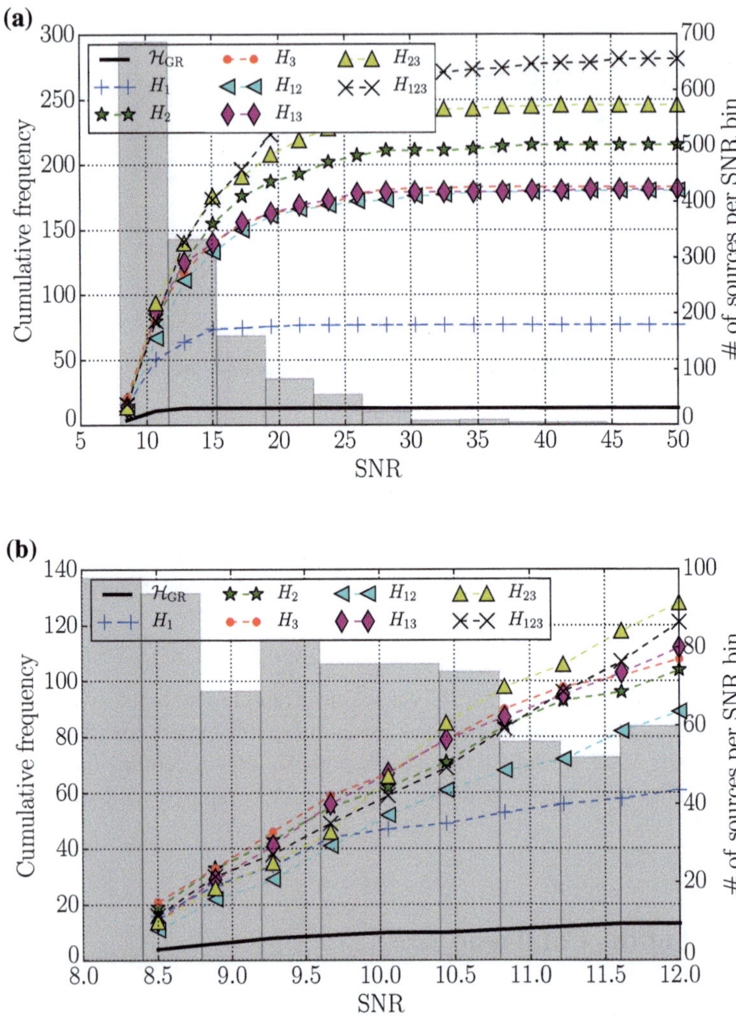

Fig. 8.19 **a** *Curves* and *left vertical axis* For a given SNR, the cumulative frequency that the Bayes factor against noise for a particular sub-hypothesis is the largest, for signals with a deviation of the form given by Eq. (8.3). All 1376 simulated sources were used. In this case, the most inclusive sub-hypothesis, H_{123}, is the most informative sub-hypothesis. The GR hypothesis rarely has the highest Bayes factor, meaning that the deviation is significant in all sub-hypotheses. *Histogram* and *right vertical axis* The number of sources per SNR bin. **b** The same as above, but restricted to sources with SNR < 12. For the low SNR sources, the sub-hypothesis H_{123} is not the best performing sub-hypothesis. Instead, only for SNR $\gtrsim 17$ does H_{123} become the most informative

As is evident from Fig. 8.21, the efficiency rises as a function of the catalogue size. This highlights the importance of combining all available sources in the advanced-detector era, when one looks for deviations from GR. The maximum catalogue size

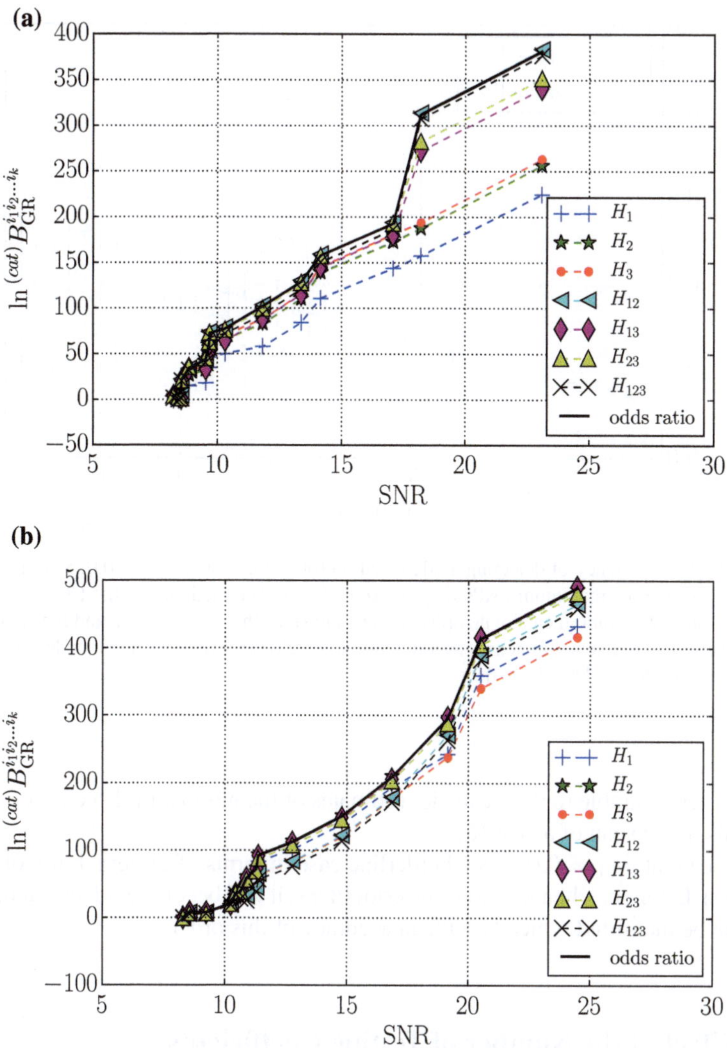

Fig. 8.20 Two examples of how cumulative Bayes factors against GR for individual sub-hypotheses and the odds ratio grow as sources with increasing SNR are being added within catalogues of 15 sources in total, for a phase shift given in Eq. (8.3). **a** An example catalogue in which the sub-hypotheses with only one free phase coefficient are less favoured than sub-hypotheses that have more than one phase coefficients free. Such a catalogue could be an indication that the true nature of the deviation is not solely within a single phase coefficient. **b** An example catalogue for which the sub-hypothesis H_{123} starts off having a lower Bayes factor than the other sub-hypotheses, due to Occam's razor. But as the SNR increases, it turns out that the sub-hypothesis H_{123} does describe the data better, and its Bayes factor increases faster than for the other sub-hypotheses. This behaviour can also be seen in Fig. 8.19

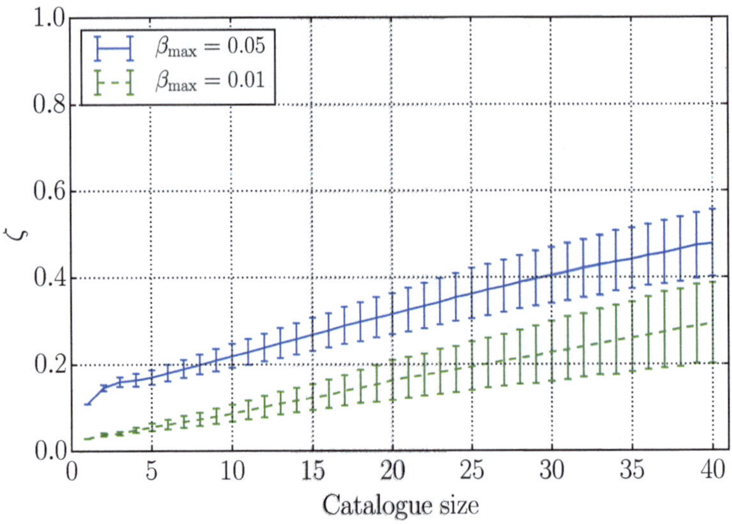

Fig. 8.21 The efficiency of detecting a GR violation for sources with $\delta\chi_3 = 0.025$, as a function of catalogue size for maximum FAPs $\beta_{\rm max} \in \{0.05, 0.01\}$. The median and the 68 % confidence interval from 5000 random catalogue orderings are shown as the *central curve* and the *error bars*, respectively. The efficiency increases as a function of catalogue size, once again underscoring the benefit of combining all available data

shown is comparable to the 'realistic' estimates of the number of detections of BNS inspirals in the span of a year [6].

We see that $\delta\chi_3 = 0.025$ is a borderline case in terms of discernability of a GR violation. Later on, when we show posterior PDFs, it will become evident that indeed, $\delta\chi_3$ can be measured typically with an accuracy of this order.

8.4 Effect of the Number of Testing Coefficients

Similar to the number of sources per catalogue, the number of sub-hypotheses is also subject to the user's choice. It is therefore of interest to see what would have happened if we had used a smaller number of testing coefficients, for example $\{\psi_1, \psi_2\}$, so that the sub-hypotheses to be tested are H_1, H_2, and H_{12}. In the example with $\delta\chi_3 = 0.025$, the PN order where the deviation occurs, namely 1.5PN, would then be higher than the PN orders associated with our testing coefficients, which are 0.5PN and 1PN.

In Fig. 8.22, the following two things are shown.

- In the case where only $\{\psi_1, \psi_2\}$ are testing coefficients, we compute the thresholds $\ln {}^{(2)}\mathcal{O}_\beta$ corresponding to maximum FAPs of $\beta_{\rm max} \in \{0.05, 0.01\}$. Next, we

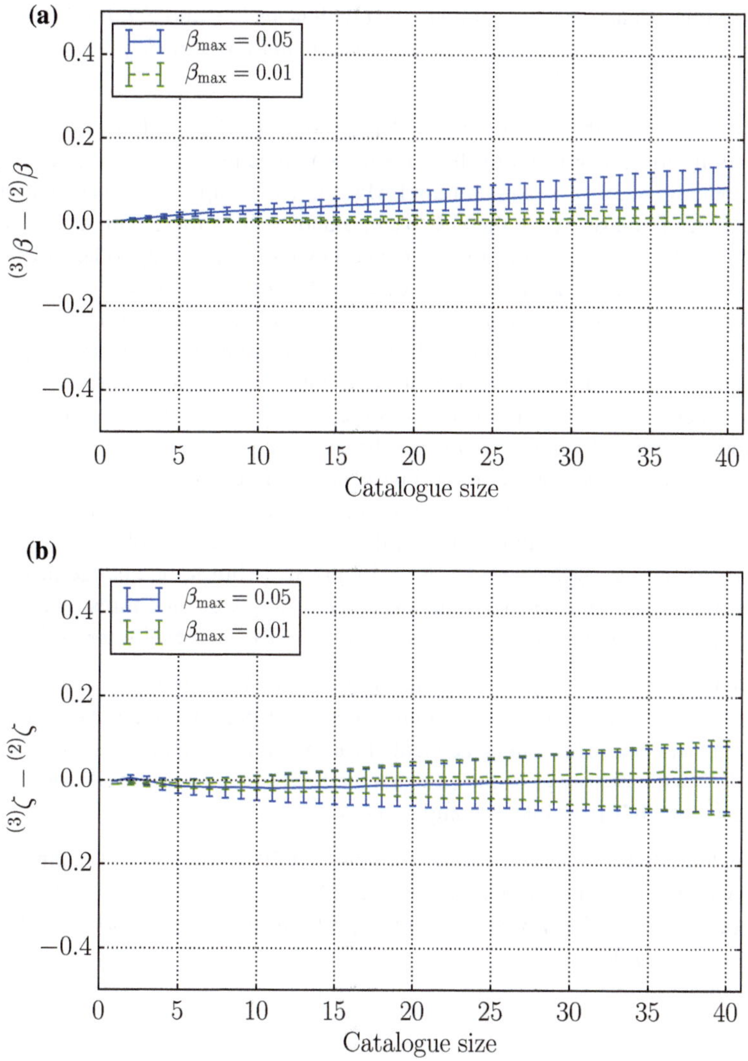

Fig. 8.22 **a** The change in FAPs in going from two to three testing coefficients, but keeping the odds ratio thresholds fixed. **b** The change in efficiencies when keeping the maximum FAPs fixed. The plots shown are for the case where the signals have $\delta\chi_3 = 0.025$. We see that increasing the number of testing coefficients has only a moderate effect on the FAPs, while for fixed maximum FAPs, the efficiencies do not change appreciably. Note, however, that when the evidence for a deviation from GR is marginal, the use of as many sub-hypotheses as possible can be pivotal in finding the violation (see Fig. 8.9, and also Fig. 8.16)

re-calculate the FAPs *for the same thresholds*, but now for the case where there are three testing coefficients, $\{\psi_1, \psi_2, \psi_3\}$, and show the difference in FAPs in Fig. 8.22a.

- On the other hand, one can compare the efficiencies ${}^{(2)}\zeta$ and ${}^{(3)}\zeta$ for the two- and three-testing-coefficients cases, *for fixed maximum FAPs* of $\beta_{\max} \in \{0.05, 0.01\}$. This is shown in Fig. 8.22b.

As expected, in the first case (fixed thresholds for the odds ratios), the FAP increases in going from two to three testing coefficients, but only moderately so. On the other hand, for fixed maximum FAPs, there is no appreciable change in efficiency. Indeed, the spread in the GR 'background' will increase with an increase in sub-hypotheses to be tested against GR; yet, having more sub-hypotheses does not really hurt us in terms of our ability to detect a deviation from GR.

Figure 8.22 indicates the typical behaviour for catalogues with a *specific* deviation from GR, in this case $\delta\chi_3 = 0.025$. It is worth repeating, however, that especially when there is only marginal evidence for a GR violation, it is important to use as many sub-hypotheses as is computationally feasible, *cf.* Fig. 8.9a. Also, we will obviously not know beforehand what the *nature* of the true GR violation is, if any.

One may nevertheless wonder how our three-testing-coefficients case would compare to a 'targeted search' that only looks for a deviation in ψ_3, which in this example happens to be where the deviation actually is. With our choice of $\alpha = 1$, this corresponds to setting $\mathcal{O}^{\text{modGR}}_{\text{GR}} = {}^{(\text{cat})}B^{3}_{\text{GR}}$. Figure 8.23 shows the change in FAPs in going from testing only H_3 to the full test for fixed log odds ratio thresholds, as well as the change in efficiencies for fixed maximum FAPs. The results are as follows.

- The change in FAPs for fixed log odds ratio thresholds is minor.
- However, especially for a large number of sources per catalogue, the efficiencies show a clear rise. This can be accounted for by the fact that, for a small violation of GR, it will not always be the case that the Bayes factor against GR for H_3 is the largest, but TIGER is able to compensate for that.

We conclude that for this *particular* example, TIGER with $N_T = 3$ testing coefficients will tend to outperform a 'targeted search' that happens to look for the violation actually present. However, we do not expect this to be true for more complicated deviations from GR.

8.5 Posteriors

Finally, let us look at some posterior PDFs for the $\delta\chi_i$. We stress that unlike Bayes factors, the PDFs cannot be combined across sources since we should not expect the $\delta\chi_i$ to be independent of the component masses; they can differ from source to source. Even looking at the PDFs for a single source may then be misleading, because even if the deviation is exactly in one or more of the PN coefficients, a given source will have values for the $\delta\chi_i$ that are representative just for the $(\mathcal{M}, \eta)$ of that source, and possibly also for the values of additional charges that may appear in an alternative theory of gravity. In a given catalogue, there may be only one source with sufficient SNR to allow for accurate parameter estimation, in which case the

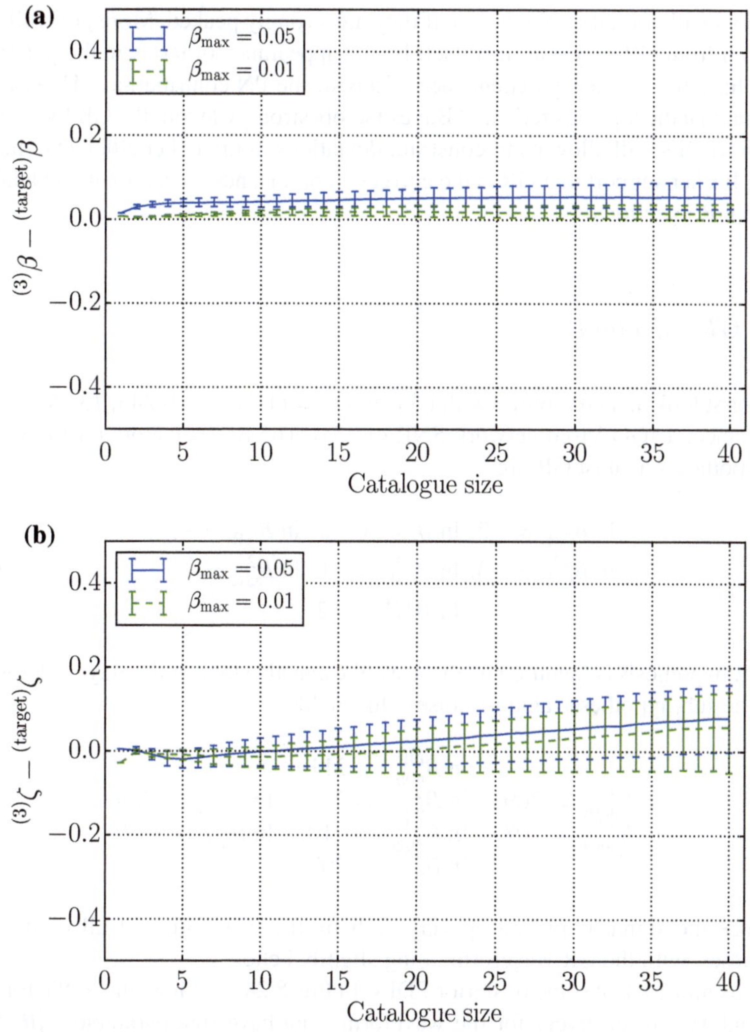

Fig. 8.23 **a** The difference of FAPs, ${}^{(3)}\beta - {}^{(\mathrm{target})}\beta$, for fixed log odds ratio thresholds and signals having $\delta\chi_3 = 0.025$, between our three-testing-coefficients case and a 'targeted search' which only looks for a deviation in ψ_3, i.e., only tests the sub-hypothesis H_3 against GR. We see that the difference is minor. **b** More important is the difference in efficiencies, ${}^{(3)}\zeta - {}^{(\mathrm{target})}\zeta$, for fixed maximum FAPs. Especially for a large number of sources per catalogue, our three-testing coefficient case is actually more efficient than the 'targeted search', at least for this particular example. This is because the Bayes factor against GR for H_3 will not be the largest in every catalogue, but TIGER naturally compensates for that. Of course, we do not expect TIGER to outperform a targeted search in the case of a more complicated deviation from GR

posteriors will not tell us much even if they are strongly peaked. More generally, the deviation from GR may manifest itself by the appearance of terms in the phase that do not have the frequency dependence of any of the PN contributions. However, in the event that the log odds ratio and Bayes factors strongly favour the GR hypothesis, posterior PDFs will allow us to constrain deviations in the PN coefficients, thereby adding further support that GR is the correct theory. Hence we start with an analysis of pure GR injections.

8.5.1 GR Injection

Let us first look at a GR source with $(\mathcal{M}, \eta, D) = (1.31\,M_\odot, 0.243, 131\,\text{Mpc})$, and an Advanced LIGO/Virgo network SNR of 23.0. The Bayes factors for the various sub-hypotheses against GR are

$$\begin{gathered} \ln B^1_{\text{GR}} = -2,\ \ln B^2_{\text{GR}} = -2,\ \ln B^3_{\text{GR}} = -2, \\ \ln B^{12}_{\text{GR}} = -3,\ \ln B^{13}_{\text{GR}} = -1,\ \ln B^{23}_{\text{GR}} = -1, \\ \ln B^{123}_{\text{GR}} = -2. \end{gathered} \tag{8.4}$$

The GR hypothesis is favoured in all cases. We can also look at the Bayes factors for all of the sub-hypotheses against noise. This yields

$$\begin{array}{ccc} & \ln B^{\text{GR}}_{\text{noise}} = 211, & \\ \ln B^1_{\text{noise}} = 209, & \ln B^2_{\text{noise}} = 209, & \ln B^3_{\text{noise}} = 209, \\ \ln B^{12}_{\text{noise}} = 208, & \ln B^{13}_{\text{noise}} = 210, & \ln B^{23}_{\text{noise}} = 210, \\ & \ln B^{123}_{\text{noise}} = 209. & \end{array} \tag{8.5}$$

Hence the signal is picked up very well by the waveforms of all of the sub-hypotheses, with the GR waveform doing slightly better.

Let us now look at some posterior PDFs. In Fig. 8.24, we show the PDFs for $\delta\chi_1$, $\delta\chi_2$, and $\delta\chi_3$, respectively for the waveforms that have free parameters $\{\boldsymbol{\theta}, \delta\chi_1\}$, $\{\boldsymbol{\theta}, \delta\chi_2\}$, and $\{\boldsymbol{\theta}, \delta\chi_3\}$, with $\boldsymbol{\theta}$ being the parameters of the GR waveform. We see that the distributions are all narrowly peaked around the correct value of zero.

8.5.2 Signal with $\delta\chi_3 = 0.1$

We now consider an example with $(\mathcal{M}, \eta, D) = (1.18\,M_\odot, 0.244, 196\,\text{Mpc})$, with a non-zero relative shift in ψ_3 of $\delta\chi_3 = 0.1$, and network SNR of 23.2. The Bayes factors in this case are

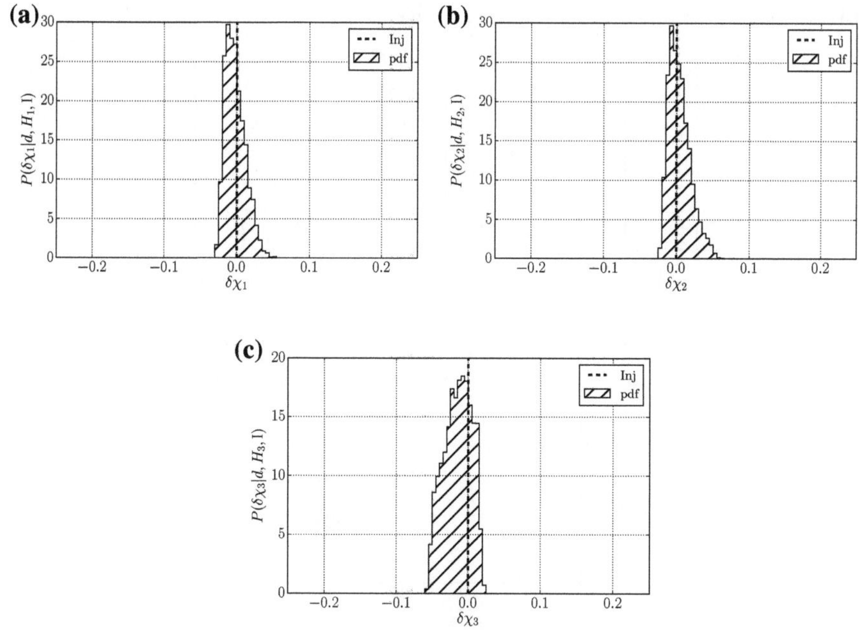

Fig. 8.24 Posterior PDFs for a GR injection. **a** $\delta\chi_1$ measured with a waveform that has $\{\boldsymbol{\theta}, \delta\chi_1\}$ as free parameters. **b** $\delta\chi_2$ measured with a waveform that has $\{\boldsymbol{\theta}, \delta\chi_2\}$ free. **c** $\delta\chi_3$ measured with a waveform that has $\{\boldsymbol{\theta}, \delta\chi_3\}$ free. In each case the distribution is tightly centred on zero, with standard deviations of 0.014, 0.015, and 0.019, respectively

$$\begin{gathered}\ln B^{1}_{\mathrm{GR}} = 117,\ \ln B^{2}_{\mathrm{GR}} = 124,\ \ln B^{3}_{\mathrm{GR}} = 124,\\ \ln B^{12}_{\mathrm{GR}} = 123,\ \ln B^{13}_{\mathrm{GR}} = 124,\ \ln B^{23}_{\mathrm{GR}} = 125,\\ \ln B^{123}_{\mathrm{GR}} = 114.\end{gathered} \tag{8.6}$$

This time the GR hypothesis is very much disfavoured. However, we note that the Bayes factor for the sub-hypothesis that only ψ_3 differs from its GR value is not the largest. In fact, all the Bayes factors except for B^{1}_{GR} and B^{123}_{GR} are rather similar in magnitude, and no clear conclusions can be drawn from them regarding the underlying nature of the deviation from GR.

When looking at the Bayes factors against noise, we see that the signal is clearly detected for all sub-hypotheses. We have

$$\begin{gathered}\ln B^{\mathrm{GR}}_{\mathrm{noise}} = 128,\\ \ln B^{1}_{\mathrm{noise}} = 245,\ \ln B^{2}_{\mathrm{noise}} = 252,\ \ln B^{3}_{\mathrm{noise}} = 252,\\ \ln B^{12}_{\mathrm{noise}} = 251,\ \ln B^{13}_{\mathrm{noise}} = 252,\ \ln B^{23}_{\mathrm{noise}} = 253,\\ \ln B^{123}_{\mathrm{noise}} = 242.\end{gathered} \tag{8.7}$$

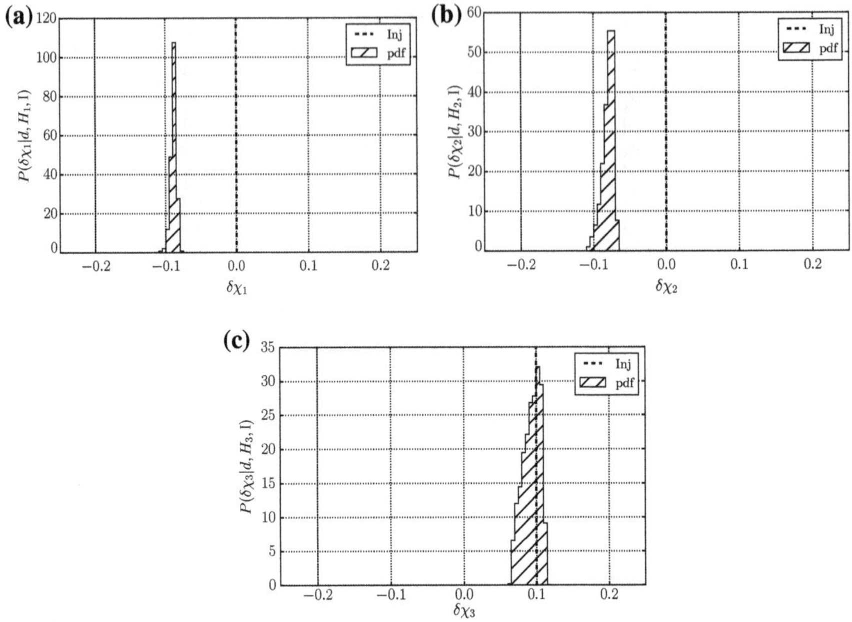

Fig. 8.25 Posterior PDFs for an injection with $\delta\chi_3 = 0.1$. **a** $\delta\chi_1$ measured with a waveform that has $\{\boldsymbol{\theta}, \delta\chi_1\}$ as free parameters; **b** $\delta\chi_2$ measured with a waveform that has $\{\boldsymbol{\theta}, \delta\chi_2\}$ free. **c** $\delta\chi_3$ measured with a waveform that has $\{\boldsymbol{\theta}, \delta\chi_3\}$ free. As expected, the PDF in panel (**c**) is sharply peaked at the correct value of $\delta\chi_3 = 0.1$, with a standard deviation of 0.012. For panels (**a**) and (**b**), the one test coefficient that is used differs from the coefficient in the signal that has the shift. The parameters in the waveform will rearrange themselves such as to best accommodate the properties of the signal. Both $\delta\chi_1$ and $\delta\chi_2$ end up being sharply peaked, but not at the correct value of zero

Now let us consider posterior PDFs. We expect the PDF of $\delta\chi_3$ for the sub-hypothesis H_3, where only $\{\boldsymbol{\theta}, \delta\chi_3\}$ are allowed to vary, to be peaked at the injected value of 0.1, and this is the case with very good accuracy, as is shown in Fig. 8.25c.

In Fig. 8.25a, b, the PDFs of $\delta\chi_1$ for the sub-hypothesis H_1, and of $\delta\chi_2$ for the sub-hypothesis H_2 are shown. In these cases, the parameter in the signal that has the shift is now not represented; in the first case only $\delta\chi_1$ is allowed to vary on top of the parameters $\boldsymbol{\theta}$ of GR, and in the second case only $\delta\chi_2$. In the Nested Sampling process, the waveform will still try to adapt itself to the deformation in the signal. The result is that $\delta\chi_1$ and $\delta\chi_2$ are strongly peaked, but away from the correct values $\delta\chi_1 = \delta\chi_2 = 0$. Thus, if one were to study the data only with waveforms from a *specific* alternative theory of gravity (e.g. a 'massive graviton' model with a deviation in ψ_2 only), one might find a violation of GR but it may lead to the wrong conclusions about the nature of the deviation.

We can also look at the PDFs for the sub-hypothesis H_{123}, where the waveforms have $\delta\chi_1$, $\delta\chi_2$, $\delta\chi_3$ free, shown in Fig. 8.26. Once again the peaks for $\delta\chi_1$, $\delta\chi_2$ and $\delta\chi_3$ are more or less at the correct value, but we now have a much larger spread.

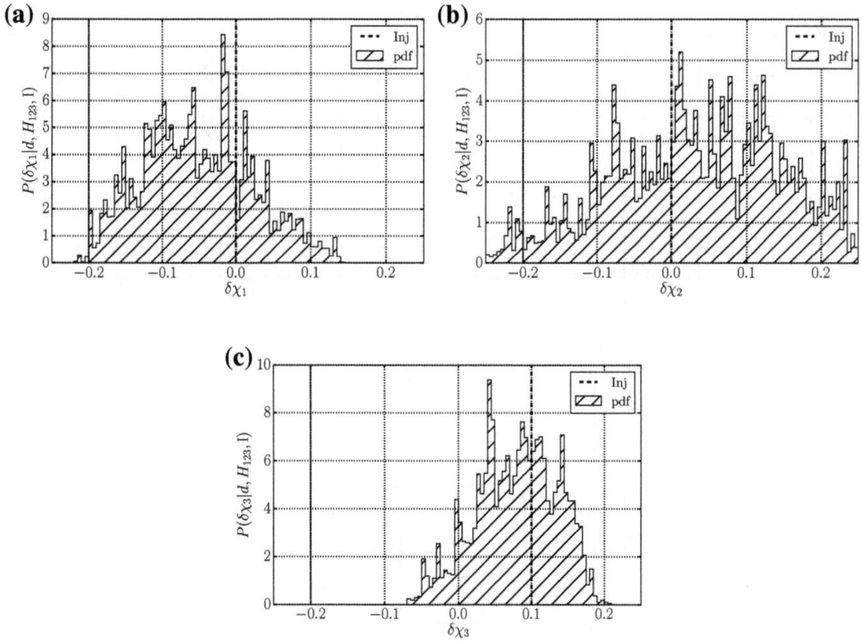

Fig. 8.26 The posterior PDFs of **a** $\delta\chi_1$, **b** $\delta\chi_2$ and **c** $\delta\chi_3$, for an injection with $\delta\chi_3 = 0.1$, recovered with waveforms where $\{\boldsymbol{\theta}, \delta\chi_1, \delta\chi_2, \delta\chi_3\}$ are all free. The peak for $\delta\chi_3$ is near the correct value of $\delta\chi_3 = 0.1$ (with a median of 0.083), but this time the spread is considerably larger (with a standard deviation of 0.055), as we are trying to measure more parameters at the same time. Both $\delta\chi_1$ and $\delta\chi_2$ are also peaked around their correct values of $\delta\chi_1 = \delta\chi_2 = 0$, but with larger spreads compared to the PDFs when only a single parameter is free

This too is as expected, as parameter estimation degrades if one tries to measure too many parameters at once.

8.5.3 Signal with $\delta\chi_3 = 0.025$

Let us consider an example with $(\mathcal{M}, \eta, D) = (1.14\, M_\odot, 0.242, 216\,\text{Mpc})$, $\delta\chi_3 = 0.025$, and a network SNR of 20.6. As expected, the Bayes factors for the sub-hypotheses against GR are considerably smaller than in the case of $\delta\chi_3 = 0.1$, but GR is still disfavoured:

$$\begin{gathered} \ln B^1_{\text{GR}} = 11,\ \ln B^2_{\text{GR}} = 12,\ \ln B^3_{\text{GR}} = 12, \\ \ln B^{12}_{\text{GR}} = 10,\ \ln B^{13}_{\text{GR}} = 11,\ \ln B^{23}_{\text{GR}} = 11, \\ \ln B^{123}_{\text{GR}} = 11. \end{gathered} \tag{8.8}$$

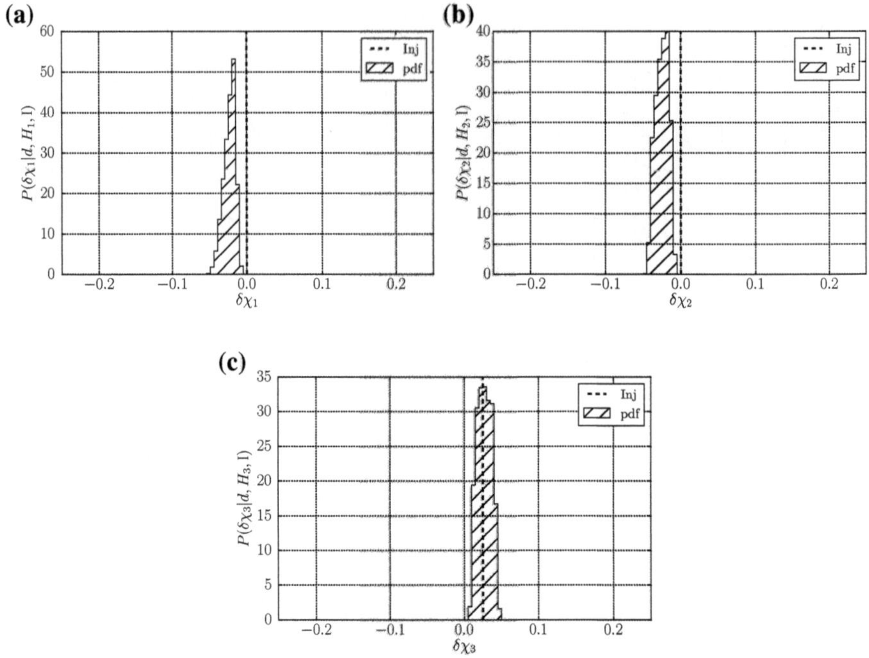

Fig. 8.27 The posterior PDFs for **a** $\delta\chi_1$, **b** $\delta\chi_2$, and **c** $\delta\chi_3$, for an injection with $\delta\chi_3 = 0.025$, recovered with waveforms where $\{\boldsymbol{\theta}, \delta\chi_1\}$, $\{\boldsymbol{\theta}, \delta\chi_2\}$, and $\{\boldsymbol{\theta}, \delta\chi_3\}$ are free, respectively. Again $\delta\chi_3$ is peaked at close to the correct value, with a median of 0.027 and a standard deviation of 0.0092, but both $\delta\chi_1$ and $\delta\chi_2$ are peaked at incorrect values

Also as expected, the signal is easily found by all of the model waveforms:

$$\begin{gathered}\ln B^{\mathrm{GR}}_{\mathrm{noise}} = 186,\\ \ln B^{1}_{\mathrm{noise}} = 197,\ \ln B^{2}_{\mathrm{noise}} = 198,\ \ln B^{3}_{\mathrm{noise}} = 198,\\ \ln B^{12}_{\mathrm{noise}} = 196,\ \ln B^{13}_{\mathrm{noise}} = 197,\ \ln B^{23}_{\mathrm{noise}} = 197,\\ \ln B^{123}_{\mathrm{noise}} = 197.\end{gathered} \tag{8.9}$$

As before, we look at the posterior PDF of $\delta\chi_3$ for the sub-hypothesis H_3, where only $\{\boldsymbol{\theta}, \delta\chi_3\}$ are allowed to vary, as shown in Fig. 8.27c. The distribution is peaked near the correct value and stays away from zero; however, one should not expect the same to happen for lower-SNR sources. Let us also look at the PDF for $\delta\chi_1$ when $\{\boldsymbol{\theta}, \delta\chi_1\}$ are free parameters, and of $\delta\chi_2$ when $\{\boldsymbol{\theta}, \delta\chi_2\}$ are free, as shown Fig. 8.27a, b. As before, $\delta\chi_1$ and $\delta\chi_2$ are not peaked at the right values of $\delta\chi_1 = \delta\chi_2 = 0$.

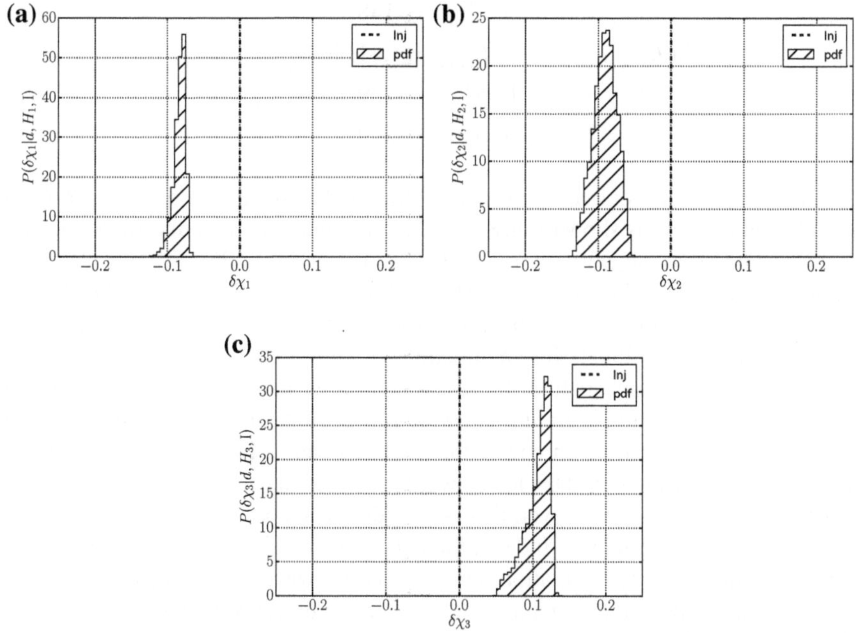

Fig. 8.28 The posterior PDFs for an injection with a deviation of the form Eq. (8.1), recovered with waveforms where, respectively, **a** $\{\boldsymbol{\theta}, \delta\chi_1\}$, **b** $\{\boldsymbol{\theta}, \delta\chi_2\}$ and **c** $\{\boldsymbol{\theta}, \delta\chi_3\}$ are free. The distribution of $\delta\chi_3$ has its median at 0.11 and a standard deviation of 0.017. Note the remarkable resemblance with Fig. 8.25, where the signal had $\delta\chi_3 = 0.1$. Also for $\delta\chi_1$ and $\delta\chi_2$, the distributions are very similar to the ones for a signal with $\delta\chi_3 = 0.1$

8.5.4 Signal with Non-PN Frequency Dependence in the Phasing

We now look at a signal with a frequency dependence between 1PN and 1.5PN, as in Eq. (8.1). In the example we use here, we have $(\mathcal{M}, \eta, D) = (1.29\,M_\odot, 0.250, 208$ Mpc), with a network SNR of 22.4. The Bayes factors for the sub-hypotheses against GR are

$$\begin{aligned} \ln B^1_{\rm GR} &= 91,\ \ln B^2_{\rm GR} = 93,\ \ln B^3_{\rm GR} = 89, \\ \ln B^{12}_{\rm GR} &= 92,\ \ln B^{13}_{\rm GR} = 91,\ \ln B^{23}_{\rm GR} = 92, \\ \ln B^{123}_{\rm GR} &= 91. \end{aligned} \tag{8.10}$$

Thus, also in this case, the GR hypothesis is very much disfavoured, despite the fact that none of our model waveforms contains the non-PN frequency dependence which is present in the phase of the signal. We can also look at the Bayes factors against noise. This yields

$$\begin{aligned}
&\ln B^{\mathrm{GR}}_{\mathrm{noise}} = 148, \\
\ln B^{1}_{\mathrm{noise}} = 239,\ &\ln B^{2}_{\mathrm{noise}} = 241,\ \ln B^{3}_{\mathrm{noise}} = 237, \\
\ln B^{12}_{\mathrm{noise}} = 240,\ &\ln B^{13}_{\mathrm{noise}} = 239,\ \ln B^{23}_{\mathrm{noise}} = 239, \\
&\ln B^{123}_{\mathrm{noise}} = 239.
\end{aligned} \tag{8.11}$$

It is interesting to look at the posterior PDF of $\delta\chi_3$ for the case where $\{\boldsymbol{\theta}, \delta\chi_3\}$ are allowed to vary, as shown Fig. 8.28c. The distribution looks uncannily like the analogous one for a signal with $\delta\chi_3 = 0.1$, shown in Fig. 8.25c. We can also look at the PDF of $\delta\chi_1$ in the case where $\{\boldsymbol{\theta}, \delta\chi_1\}$ are free parameters, and the PDF of $\delta\chi_2$ when $\{\boldsymbol{\theta}, \delta\chi_2\}$ are free, as shown in Fig. 8.28a, b. Here too there is an interesting resemblance to Fig. 8.25a, b, for an injection with $\delta\chi_3 = 0.1$.

In conclusion, trying to find the specific form of the deviation by looking at the posterior distributions can ultimately be misleading. Indeed, the posteriors can 'masquerade' as deviations in one of the PN coefficients. Now, one could imagine looking for patterns in the PDFs to see what information can be extracted about the nature of the deviation. However, this is outside the scope of this work and would also be somewhat premature. Indeed, the behaviour would not only depend on the precise nature of the GR violation, but also on the waveform used. In the future, we may want to use time domain waveforms, e.g. numerically evolved by using a Hamiltonian, and then construct model waveforms by introducing deviations directly in that Hamiltonian. Whatever patterns may emerge depending on the GR violation will then pertain to the associated coefficients, which would be more directly related to the physics of coalescence than the phasing coefficients. We leave this issue for future work.

References

1. C.K. Mishra et al., Parametrized tests of post-Newtonian theory using advanced LIGO and Einstein telescope. Phys. Rev. D **82**, 064010 (2010)
2. R. O'Shaughnessy, C. Kim, Pulsar binary birth rates with spin-opening angle correlations. Astrophys. J. **715**, 230–241 (2010)
3. C. Van Den Broeck, A.S. Sengupta, Binary black hole spectroscopy. Class. Quantum Gravity **24**(5), 1089 (2007)
4. C. Van Den Broeck, A.S. Sengupta, Phenomenology of amplitude-corrected post-Newtonian gravitational waveforms for compact binary inspiral: I. Signal-to noise ratios. Class. Quantum Gravity **24**(1), 155 (2007)
5. S. Vitale, M. Zanolin, Parameter estimation from gravitational waves generated by nonspinning binary black holes with laser interferometers: beyond the fisher information. Phys. Rev. D **82.12**, 124065 (2010)
6. J. Abadie et al., Calibration of the LIGO gravitational wave detectors in the fifth science run. Nucl. Instrum. Methods Phys. Res. A **624**, 223–240 (2010)
7. F. Özel et al., On the mass distribution and birth masses of neutron stars. ApJ **757**, 55 (2012)
8. J. Abadie et al., Predictions for the rates of compact binary coalescences observable by ground-based gravitational-wave detectors. Class. Quantum Gravity **27**(17), 173001 (2010)
9. C. Cutler et al., The last three minutes—issues in gravitational-wave measurements of coalescing compact binaries. Phys. Rev. Lett. **70**, 2984–2987 (1993)

10. A. Buonanno et al., Comparison of post-Newtonian templates for compact binary inspiral signals in gravitational-wave detectors. Phys. Rev. D **80**, 084043 (2009)
11. J. Veitch, A. Vecchio, Bayesian coherent analysis of in-spiral gravitational wave signals with a detector network. Phys. Rev. D **81**, 062003 (2010)
12. M. Dominik et al., Double compact objects. I. The significance of the common envelope on merger rates. ApJ **759**, 52 (2012)
13. J. Abadie et al., Search for gravitational waves from compact binary coalescence in LIGO and Virgo data from S5 and VSR1. Phys. Rev. D **82.10**, 102001 (2010)
14. L. Blanchet, B.S. Sathyaprakash, Signal analysis of gravitational wave tails. Class. Quantum Gravity **11**(11), 2807 (1994)
15. L. Blanchet, B.S. Sathyaprakash, Detecting a tail effect in gravitational-wave experiments. Phys. Rev. Lett. **74**, 1067–1070 (1995)
16. L.C. Stein, N. Yunes, Effective gravitational wave stress-energy tensor in alternative theories of gravity. Phys. Rev. D **83**, 064038 (2010)
17. K. Yagi et al., Post-Newtonian, quasicircular binary inspirals in quadratic modified gravity. Phys. Rev. D **85.6**, 064022 (2012)
18. N. Yunes, L.C. Stein, Non-spinning black holes in alternative theories of gravity. Phys. Rev. D **83**, 104002 (2011)
19. N. Yunes, S.A. Hughes, Binary pulsar constraints on the parametrized post-Einsteinian framework. Phys. Rev. D **82**, 082002 (2010)

Chapter 9
Discussion

We have developed TIGER, a general framework to search for deviations from GR by using signals from CBC events. To this end, we constructed an odds ratio $O_{\mathrm{GR}}^{\mathrm{modGR}}$, which is the posterior probability that there is a deviation from GR, versus GR being correct. This odds ratio can be written as a linear combination of Bayes factors $B_{\mathrm{GR}}^{i_1 i_2 \ldots i_k}$ for sub-hypotheses $H_{i_1 i_2 \ldots i_k}$, in each of which one or more of the phase parameters ψ_i is assumed to deviate from the GR value, without actually assuming any specific dependence on the frequency and/or physical parameters pertinent to a given theory. Since this includes sub-hypotheses where only a single one of the ψ_i is non-GR, our method will be particularly well-suited in low-SNR scenarios, which we expect to be relevant for the upcoming advanced-detector network. We also showed that information from multiple sources can be combined to arrive at an odds ratio $\mathcal{O}_{\mathrm{GR}}^{\mathrm{modGR}}$ for the 'catalogue' of all observed events. Furthermore, we showed that an observed odds ratio can be interpreted through the comparison with a distribution of the odds ratio originating from simulated GR signals embedded in noise, the so-called 'background' distribution. A measured odds ratio can then be assigned a 'false alarm probability', quantifying our confidence in a deviation. Finally, in the absence of real detections, we show how one can assess the probability of detecting a specific deviation from GR. This is done through the construction of the foreground, which is the distribution of the odds ratio when the signals are given a specific deviation and are embedded in noise. The quantity 'efficiency' is subsequently defined to quantify the fraction of the foreground above a certain threshold, set by a given false alarm probability.

In order to gauge the performance of the method, we have analysed two classes of deviations. The first class involved deviations which are in one of the phase coefficients. In the second category of deviations, we move away from this restriction and allow the deviation to manifest itself in arbitrary locations. Although these examples were heuristic, they allowed us to investigate the performance of the method for waveforms with deviations from GR that cannot be accounted for by a simple shift in one of the phase coefficients.

T.G.F. Li, *Extracting Physics from Gravitational Waves*,
Springer Theses, DOI 10.1007/978-3-319-19273-4_9

We first considered signals with a constant fractional deviation in the 1.5PN coefficient ψ_3. This coefficient is of particular interest, since it incorporates the so-called 'tail effects' [1, 2] (as well as spin-orbit coupling [3], although we did not consider spin here), which are not empirically accessible with binary pulsar observations and can only be studied through direct detection of gravitational waves. When considering catalogues of only 15 BNS sources, we saw that a deviation in ψ_3 at the 10 % level would easily be detectable. In fact, even a deviation at the few per cent level can be discernible. This is confirmed by posterior PDFs for ψ_3 in the case where this is the only parameter that is assumed to deviate from its GR value.

We further showed results for signals with a shift in the 2PN phase coefficient, ψ_4. Setting $\delta\chi_4 = 0.2$, the induced change in phase at $f = 150$ Hz is comparable to a constant shift $\delta\chi_3 = -0.035$ in the 1.5PN coefficient. The choice of a modification at the 2PN coefficient was inspired by corrections to the phase if one considers a modified Einstein-Hilbert action containing terms that are quadratic in the Riemann tensor [4–6]. As we have shown, direct detection of GWs analysed with TIGER can constrain theories of this kind to precisions that cannot be achieved by binary pulsar observations.

We also considered a deviation in the phase with a frequency dependence that does not match any of the PN coefficients, and hence is not present in any of the model waveforms that we used. More precisely, we looked at signals whose phase has an additional contribution, which has a frequency dependence in between that of the 1 and 1.5PN terms ('1.25PN'). The magnitude of the deviation was chosen such that near $f \sim 150$ Hz, where the detectors are the most sensitive, the change in phase is roughly the same as the change caused by a 10 % shift in the 1.5PN coefficient. The deviation was clearly detectable in the log odds ratios, the Bayes factors, and the posterior PDFs. We expect this to be an instance of a more general fact. Namely, even if there is a deviation in the phase which the model waveforms technically do not allow for, it will typically be observable, on condition that near the sensitive region of the detector, it causes a change in the phase that is comparable to the effect of a shift in the (low order) PN phase coefficients of more than a few per cent.

In the last example, signals were studied that have a deviation in the phase with a mass dependent power of frequency, effectively ranging from 0.5 to 1.5PN as the total mass is varied from the lowest to the highest value we consider. The magnitude of the effect was such that at $f = 150$ Hz, the change in phase was the same as that induced by a constant relative shift $\delta\chi_3 = 0.1$. One concludes that even a fully mass dependent deviation (i.e. in both the magnitude and the frequency dependence) will be detectable with our method. This strengthens our belief that any deviation, regardless of its form, can be found through TIGER, as long as the induced phase shift is of a detectable magnitude.

9.1 Comparison with Existing Methods

Having seen the results in Chap. 8, it is instructive to compare and contrast our findings with the tests of GR from the existing literature (see Sect. 6.4).

A direct comparison with the proposal by Arun et al. can only be made at the level of the posteriors. Before doing so, it should be stressed again that TIGER is not based on the posteriors but on the odds ratio. Posteriors in Sect. 8.5 are merely shown as a sanity check and to argue that a method based on posteriors can lead to a severe misconception about the underlying nature of the deviation from GR.

As shown in Figs. 8.25, 8.26 and 8.27, the precision of the measurements of ψ_3 is in the order of 10^{-2}. This is in accordance with results from Fisher matrix calculations. In addition, our results show that, as a by-product, the systematic effects are negligible compared to the statistical errors. One can thus conclude that the proposal by Arun et al. can be successfully implemented.

Nevertheless, the method by Arun et al. still suffers from a major drawback. By looking at posteriors, it is only possible in limited cases to combine information from multiple sources. Consider a theory that modifies some phase coefficient, ψ_i, but does this in a non-constant way. A simple example would be that the deviation depends on the compactness of the system and therefore ultimately depends on the masses of the system. Measurements of the ψ_i parameter can then not be combined across sources with different masses. The exception is when one knows what the dependence of the shift is in terms of the masses. However, this would go against the generality that TIGER seeks. Furthermore, one has to compute the predicted functional forms and analyse the data for each alternative theory.

Further comparison on the parameter estimation side can be done with the work by Del Pozzo et al. [7]. In this work, the authors considered a direct search in a Bayesian setting for the massive graviton with a simplified waveform derived by Will [8]. Del Pozzo et al. showed that the Compton wavelength, λ_g, can be bounded to $\lambda_g > 6 \times 10^{15}$ m, which is a slight improvement over the $\lambda_g > 2.8 \times 10^{15}$ m derived from solar system tests [8]. Such a bound can be compared to Fig. 8.24, in which we also analyse the posteriors for the case in which the signal was in accordance with GR. Assuming the simplified waveform as shown by Will [8], the measurement of $\delta\chi_2$ can be directly translated into the Compton wavelength, as both occur as a shift in the 1.0PN phase coefficient. As shown in Fig. 8.24, the limit on the massive graviton from a single measurement is about $\delta\chi_2 \sim 0.05$, which translates into $\lambda_g > 3 \times 10^{15}$ m. Although this example only considers a single instance, it is evident that the resulting posteriors on $\delta\chi_2$ are comparable to the results of Del Pozzo et al., increasing our confidence in the implementation of TIGER.

A comparison can also be made with work related to PPE [4–6, 9–14]. As the main goal of PPE is to pinpoint the nature of the deviation from GR through the inference of extra free parameters, this approach suffers, like any approach that considers explicit deviations, from severe drawbacks. Firstly, parameter estimation significantly degrades if one includes many free coefficients. Secondly, the model waveform needs to be able to account for the deviation exactly, in order to avoid being misinformed by systematic errors. An example of this can be seen by comparing Figs. 8.25 to 8.28. In this example, we show that a deviation that cannot be accounted for by the model waveform can still behave as if it were a deviation that is accounted for by the model waveform.

However, in their recent work, Cornish et al. [15] implemented the PPE waveform into a Bayesian analysis. This approach allows for Bayes factors to be calculated and therefore the PPE hypothesis to be compared to the GR hypothesis. A few words of caution are in place before we compare their results to ours. Firstly, the calculations in Cornish et al. were performed in the absence of noise. As can be seen in Fig. 8.1, noise can indeed produce a non-negligible odds ratio and a proper treatment of noise ought to be performed along the lines of Sect. 7.2. Furthermore, Cornish et al. consider sources with an SNR of 20. Such an event would be considered relatively rare in the advanced-detector era, as most sources are expected to be near threshold (SNR = 8). Our work considers a more realistic (albeit speculative) noise contribution and the SNR of sources are taken as low as the threshold for a detection claim. As the specific form of the deviation was not considered in Chap. 8, we instead compare the odds ratio for a similar shift in magnitude of the phase induced by the deviation at $f = 150$ Hz. Despite performing under more challenging circumstances (including noise and considering sources down to SNR of 8), the results are found to be comparable. This gives us confidence that TIGER is more efficient than other generic tests applicable to CBC signals.

9.2 Outlook

The preliminary results presented here motivate the construction of a full data analysis pipeline for testing or constraining GR. A test of GR on Advanced LIGO/Virgo data could be implemented along the following lines.

- Starting from particular GR waveforms, introduce parameterised deformations, leading to disjoint sub-hypotheses like our $H_{i_1 i_2 \ldots i_k}$, which together form $\mathcal{H}_{\mathrm{modGR}}$.
- Use many injections of GR waveforms in real or realistic data, to investigate the background distribution of the odds ratio, $P\left(\ln \mathcal{O}_{\mathrm{GR}}^{\mathrm{modGR}} | \kappa, \mathcal{H}_{\mathrm{GR}}, \mathrm{I}\right)$. Use the background together with choices for the maximum false alarm probability to set *thresholds*, $\ln \mathcal{O}_\beta$, for the *measured* odds ratio and Bayes factors to overcome.
- Apply TIGER to the catalogue of sources actually found by the detectors. If the measured odds ratio $\mathcal{O}_{\mathrm{GR}}^{\mathrm{modGR}}$ is below threshold, then there is no real reason to believe that a deviation from GR is present. The posterior PDFs for the free phase and amplitude coefficients in the model waveforms, taken from the highest-SNR sources, could provide (potentially very strong) constraints on these parameters.
- If the measured odds ratio is above threshold, then a violation of GR is likely. As we have seen, Bayes factors and PDFs can be misleading in trying to find out what the precise nature of the deviation may be. However, one may be able to follow up on the violation by again using TIGER, this time with waveforms with

more complicated deformations and a larger number of free parameters, inspired by particular alternative theories of gravity, similar to what is done in PPE [9].

Waveform Approximant

We note that TIGER is not tied to any particular waveform approximant. Moreover, the deformations need not be in the phase. Indeed, in the future one would presumably want to use time domain waveforms, for which it may be more convenient (and physically more appropriate) to introduce parameterised deformations directly in coefficients appearing in, e.g., a Hamiltonian used to evolve the inspiral part of the waveform. Irrespective of the parameterisation, one would still be able to associate with it an exhaustive set of logically disjoint sub-hypotheses $H_{i_1 i_2 \ldots i_k}$, and calculate the corresponding odds ratio.

For BNS systems, the `TaylorF2` approximation might turn out to be sufficiently accurate in the inspiral regime to find deviations from GR, as it seems to have a good match with numerical simulations in the inspiral regime [16]. Although neutron stars are extended objects and the `TaylorF2` waveform is only valid in the point-object limit, calculations on the corrections from finite-size effects suggest that up to $f = 400$ Hz, these contributions can be neglected [17]. Therefore, by only testing the inspiral regime up to a given frequency, even the finite-size effects can be ignored. In addition, considering that spin corrections [18] and higher order contributions to the amplitude [19, 20] can be neglected in the case of BNS systems and advanced detectors, the analyses shown in this work can be viewed as the forerunner of a future data analysis pipeline.

Sources

Finally, the theoretical concepts of TIGER are not bound to BNS systems, but can be applied on a much broader range of sources, such as BBH systems or BHNS systems. Indeed, the extension of the present work to include a broader range of sources is part of an ongoing effort by the author and his collaborators. For these sources, one will need to use TIGER with more sophisticated GR waveforms compared to the `TaylorF2` waveforms. Such waveforms ought to include as many effects as possible that GR predicts, such as merger and ringdown, higher harmonics both in the inspiral and ringdown parts, dynamical spins, and residual eccentricity. The development of such waveforms, with input from numerical simulations, is currently a subject of intense investigation [21–31].

One of the challenges of extending TIGER to BHNS and BBH is to include spin. Although spinning waveforms do exist (e.g.[27, 30, 31]), they are computationally too expensive to analyse thousands of simulated sources required to construct a background (see Sects. 7.2 and 8.2.1). Speeding up waveform calculations and computational advances are therefore necessary before TIGER can be applied to BHNS and BBH sources.

Another challenge is that for BHNS and BBH systems, the merger and ringdown signals are in the sensitive region of the detector (the last stable orbit occurs at $f \sim 400$ Hz for BHNS systems and $f \sim 200$ Hz for BBH systems). Although some phenomenological waveforms model the entire inspiral, merger and ringdown signal (e.g. [21–23, 29]), their correspondence to GR is yet not sufficient to use these waveforms to perform tests of GR. Waveforms that can model the entire evolution of binary systems are therefore needed before TIGER can be applied to BHNS and BBH systems.

Finally, although much work remains to be done to ensure readiness at the start of the advanced-detector era, the work presented in this book shows the skeleton of a very general method for testing GR with CBC events to be detected by the upcoming Advanced LIGO and Advanced Virgo observatories.

References

1. L. Blanchet, B.S. Sathyaprakash, Signal analysis of gravitational wave tails. Class. Quantum Gravity **11**(11), 2807 (1994)
2. L. Blanchet, B.S. Sathyaprakash, Detecting a tail effect in gravitational-wave experiments. Phys. Rev. Lett. **74**, 1067–1070 (1995)
3. L. Blanchet et al., Gravitational-radiation damping of compact binary systems to second post-Newtonian order. Phys. Rev. Lett. **74**, 3515–3518 (1995)
4. L.C. Stein, N. Yunes, Effective gravitational wave stress-energy tensor in alternative theories of gravity. Phys. Rev. D **83**, 064038 (2010)
5. K. Yagi et al., Post-Newtonian, quasicircular binary inspirals in quadratic modified gravity. Phys. Rev. D **85**(6), 064022 (2012)
6. N. Yunes, L.C. Stein, Non-spinning black holes in alternative theories of gravity. Phys. Rev. D **83**, 104002 (2011)
7. W. Del Pozzo, J. Veitch, A. Vecchio, Testing general relativity using Bayesian model selection: applications to observations of gravitational waves from compact binary systems. Phys. Rev. D **83**, 082002 (2011)
8. C.M. Will, Bounding the mass of the graviton using gravitational-wave observations of inspiralling compact binaries. Phys. Rev. D **57**, 2061–2068 (1998)
9. N. Yunes, F. Pretorius, Fundamental theoretical bias in gravitational wave astrophysics and the parametrized post-Einsteinian framework. Phys. Rev. D **80**, 122003 (2009)
10. N. Yunes, S.A. Hughes, Binary pulsar constraints on the parametrized post- Einsteinian framework. Phys. Rev. D **82**, 082002 (2010)
11. N. Yunes, R. O'Shaughnessy, B.J. Owen, S. Alexander, Testing gravitational parity violation with coincident gravitational waves and short gamma ray bursts. Phys. Rev. D **82**, 064017 (2010)
12. N. Yunes, F. Pretorius, D. Spergel, Constraining the evolutionary history of Newton's constant with gravitational wave observations. Phys. Rev. D **81**, 064018 (2010)
13. S. Mirshekari, N. Yunes, C.M. Will, Constraining Lorentz-violating, modified dispersion relations with gravitational waves. Phys. Rev. D **85**(2), 024041 (2012)
14. K. Chatziioannou, N. Yunes, N. Cornish, Model-independent test of general relativity: an extended post-Einsteinian framework with complete polarization content. Phys. Rev. D **86**(2), 022004 (2012)
15. N. Cornish et al., Gravitational wave tests of general relativity with the parameterized post-Einsteinian framework. Phys. Rev. D **84**(6), 062003 (2011)
16. A. Buonanno et al., Comparison of post-Newtonian templates for compact binary inspirals ignals in gravitational-wave detectors. Phys. Rev. D **80**, 084043 (2009)
17. T. Hinderer et al., Tidal deformability of neutron stars with realistic equations of state and their gravitational wave signatures in binary inspiral. Phys. Rev. D **81**(12), 123016 (2010)
18. R. O'Shaughnessy, C. Kim, Pulsar binary birth rates with spin-opening angle correlations. Astrophys. J. **715**, 230–241 (2010)

19. C. Van Den Broeck, A.S. Sengupta, Binary black hole spectroscopy. Class. Quantum Gravity **24**(5), 1089 (2007)
20. C. Van Den Broeck, A.S. Sengupta, Phenomenology of amplitude-corrected post-Newtonian gravitational waveforms for compact binary inspiral: I. Signal-to-noise ratios'. Class. Quantum Gravity **24**(1), 155 (2007)
21. P. Ajith et al., A phenomenological template family for black-hole coalescence waveforms. Class. Quantum Gravity **24**(19), S689 (2007)
22. P. Ajith et al., Inspiral-merger-ringdown waveforms for black-hole binaries with nonprecessing spins. Phys. Rev. Lett. **106**(24), 241101 (2011)
23. P. Ajith et al., Inspiral-merger-ringdown waveforms for black-hole binaries with nonprecessing spins. Phys. Rev. Lett. **106**, 241101 (2011)
24. A. Buonanno, T. Damour, Effective one-body approach to general relativistic two-body dynamics. Phys. Rev. D **59**, 084006 (1999)
25. A. Buonanno, T. Damour, Transition from inspiral to plunge in binary black hole coalescences. Phys. Rev. D **62**, 064015 (2000)
26. T. Damour, P. Jaranowski, G. Schafer, Determination of the last stable orbit for circular general relativistic binaries at the third post-Newtonian approximation. Phys. Rev. D **62**, 084011 (2000)
27. T. Damour, Coalescence of two spinning black holes: an effective one-body approach. Phys. Rev. D **64**, 124013 (2001)
28. A. Buonanno, Y. Chen, T. Damour, Transition from inspiral to plunge in precessing binaries of spinning black holes. Phys. Rev. D **74**, 104005 (2006)
29. Y. Pan et al., Inspiral-merger-ringdown multipolar waveforms of nonspinning blackhole binaries using the effective-one-body formalism. Phys. Rev. D **84**(12), 124052 (2011)
30. R. Sturani et al., Phenomenological gravitational waveforms from spinning coalescing binaries. ArXiv e-prints (2010), arXiv:1012.5172
31. R Sturani et al., Complete phenomenological gravitational waveforms from spinning coalescing binaries. J. Phys.: Conf. Ser. **243**(1) (2010)

Part III
Inferring the Large-Scale Structure of the Universe

Chapter 10
Introduction

In part II, we used GR to describe the high compactness regime, where the typical distance is small compared to the total rest mass of the system. Assuming a uniform distribution of stellar objects, the total rest mass scales as r^3, where r is some distance. Therefore, as r increases, one must reach a point where the compactness GM/c^2r is high again. This high compactness regime is described by the field of *cosmology*, i.e. the study of the large scale behaviour of the Universe.

To be a physical description of the large scale behaviour of the Universe, GR must be able to explain current cosmological observations. A few of these observations are key to our understanding of cosmology. In 1964, Penzias and Wilson discovered a thermal radiation with a near uniform temperature coming from all directions of the sky, commonly known as the cosmic microwave background (CMB) [1]. This remarkable uniformity suggests that the Universe is *isotropic* on large scales, i.e. uniform in all spatial directions. Subsequent measurements of the CMB have all confirmed the isotropy of the CMB temperature [2, 3].

Observations of large clusters of stars, known as galaxies or nebulae, and even large clusters of galaxies suggest that the Universe contains well defined structures. However, as we observe the galaxy or galaxy cluster distributions on even larger scales ($\gtrsim$150 Mpc), we find that the Universe is remarkably *homogeneous*, i.e. it has uniform physical conditions at any position [4].

Finally, both Lemaître in 1927 and Hubble in 1929 found that all galaxies seem to be moving away from Earth. Moreover, the radial velocity with which galaxies appear to be receding from us is proportional to the distance to the galaxy [5, 6]. This behaviour is now referred to as *Hubble's law*, and suggests that the Universe is uniformly *expanding*.

T.G.F. Li, *Extracting Physics from Gravitational Waves*,
Springer Theses, DOI 10.1007/978-3-319-19273-4_10

10.1 Content of the Universe

On a cosmological scale we can view galaxies and galaxy clusters as mere particles in a gas. We further assume a negligible interaction between the 'particles' such that the content of the Universe can be modelled by a perfect fluid. A perfect fluid can be described by a 4-velocity u^μ (which is the 4-velocity of an observer who sees the fluid with no mean motion), a total energy density ρ_{tot} (the energy density as seen by an observer with a 4-velocity u^μ) and the total pressure p_{tot} (the kinetic pressure of the 'particles' perpendicular to the boundary of the fluid). This 4-velocity is the same as the mean 4-velocity of the nearby galaxies, and is the 4-velocity necessary to observe an isotropic CMB. Consequently, the stress-energy tensor describing the fluid consisting of galaxies, globular clusters, etc.—also known as the "cosmological fluid"—can be written as

$$T_{\mu\nu} = (\rho_{\text{tot}} + p_{\text{tot}})\, u_\mu u_\nu + p_{\text{tot}} g_{\mu\nu}, \tag{10.1}$$

where $g_{\mu\nu}$ is the metric tensor. There are several contributions to the total content of the Universe, each with their own characteristic behaviour. We will list the most widely accepted contributions.

Matter
In this context, matter refers to particles that have non-relativistic velocities. Matter therefore has negligible pressure compared to the energy density and can be approximated to have

$$p_M \approx 0. \tag{10.2}$$

Matter in this category includes baryonic matter and cold dark matter. The energy density of matter is denoted by ρ_M.

Radiation
Radiation comprises all massless (e.g. photons) or relativistic particles (e.g. neutrinos). For photons, the stress-energy tensor is given by

$$T^{\mu\nu} = \frac{1}{\mu_0}\left(F^{\mu\alpha}F^{\nu\alpha} - \frac{1}{4}g^{\mu\nu}F_{\alpha\beta}F^{\alpha\beta}\right), \tag{10.3}$$

where $F_{\mu\nu} = \partial_\mu A_\nu - \partial_\nu A_\mu$ is the electromagnetic field, A_μ is the electromagnetic potential, and μ_0 is the permeability of free space. From Eq. (10.3), it follows that the pressure is related to the energy density through

$$p_R = \frac{1}{3}\rho_R, \tag{10.4}$$

where energy density for radiation is denoted by ρ_R. The result in Eq. (10.4) holds for all massless particles, and is approximately true for relativistic particles.

Dark Energy
Dark energy is a hypothetical energy source which behaves differently than ordinary matter or radiation does: it has a positive energy density but a negative pressure. Dark energy is postulated to describe the accelerated expansion of the Universe, as inferred from type SNIa supernovae observations [7, 8]. As the characteristics of this hypothetical energy form are unknown, a first attempt is to model it as a perfect fluid with an equation of state given by

$$p_{DE} = w\rho_{DE}. \tag{10.5}$$

To account for the accelerated expansion of the Universe, the equation of state parameter has to be $w < 0$. As it turns out, dark energy is equivalent to having a cosmological constant in the EFE (*cf* Sect. 10.3) for the value $w = -1$. In general, the equation of state parameter can be a function of time, corresponding to an evolving equation of state.

10.2 Topology of the Universe

Given the observational evidences for isotropy, homogeneity, and uniform expansion, a natural starting point to describe the Universe on large scales is to postulate isotropy, homogeneity and uniform expansion. These postulates are in agreement with the Copernican principle, i.e. humans are not privileged observers of the Universe.

Before we describe the Universe using GR, we must first clearly define the meanings of homogeneity and isotropy. Homogeneity of the Universe means that through each event in the Universe there passes a spacelike hypersurface on which the physical conditions (e.g. energy density, pressure, curvature) are identical. Isotropy of the Universe means that an observer who moves with the cosmological fluid cannot distinguish any preferred spatial direction by any local measurements. Therefore, isotropy guarantees that the world line of any observer that moves with the cosmological fluid is orthogonal to the hypersurface of homogeneity.

10.2.1 Comoving Coordinate System

Next, we want to write down a generic metric that satisfies all these postulates, by using a coordinate system that make these properties apparent. Suppose we have a hypersurface of homogeneity, denoted by S_I. On this spacelike hypersurface, we impose a spatial coordinate system, where each event is assigned the spatial coordinate (x^1, x^2, x^3) and we assign a specific coordinate time, denoted by t_I. Then, we allow this hypersurface of homogeneity to evolve throughout spacetime by letting each observer on this hypersurface to move with the cosmological fluid. However, we impose that all events on the world line of a specific observer that intersects the surface S_I at (x'^1, x'^2, x'^3) remains on the same spatial coordinates (x'^1, x'^2, x'^3). Therefore, an observer that moves with the cosmological fluid is at rest with respect to these coordinates, which are commonly referred to as *comoving coordinates*. The time coordinate can be defined according to the lapse of proper time between event P_i on the hypersurface S_I and an event P_e on the world line of an observer that moves with the cosmological fluid, plus the coordinate time associated to the surface S_I, denoted by t_I. In other words, the comoving coordinate time can be defined as

$$t = t_I + \int_{\tau(P_i)}^{\tau(P_e)} d\tau, \tag{10.6}$$

where τ is the proper time.

For an event P, the basis vectors $\partial/\partial x^i$ are tangent to the hypersurface of homogeneity that goes through P and that the basis vector $\partial/\partial t$ is tangent to the world line going through P. Therefore, $\partial/\partial t$ is orthogonal to $\partial/\partial x^i$, i.e.

$$\frac{\partial}{\partial t} \cdot \frac{\partial}{\partial x^i} = 0 \quad \text{for} \quad i = 1, 2, 3. \tag{10.7}$$

By noting that the comoving coordinate time also measures the lapse of proper time, we can write $\partial/\partial t = u$, where u denotes the 4-velocity of the cosmological fluid, so that

$$\partial/\partial t \cdot \partial/\partial t = -1. \tag{10.8}$$

Using Eqs. (10.7) and (10.8), we can therefore write the line element as

$$ds^2 = -dt^2 + g_{ij} dx^i dx^j. \tag{10.9}$$

10.2.2 Hypersurface of Homogeneity

Next, we take a closer look at the spatial components of the metric, g_{ij}. In particular, we look at how the 3-metric of hypersurfaces of homogeneity evolve from an arbitrary initial hypersurface, and at the nature a hypersurface of homogeneity.

Suppose that two adjacent world lines of observers that move along the cosmological fluid are separated by the proper distance

$$\Delta\sigma(t_I) = \sqrt{g_{ij}(t_I, x^k)\Delta x^i \Delta x^j}, \tag{10.10}$$

where t_I is the initial co-moving time, $g_{ij}(t_I, x^k)$ the metric at time t_I and spatial position x^k, and Δx^i is the coordinate separation in the ith direction. At some later time t, the spatial separation is given by $\Delta\sigma(t)$, so that the fractional change of the proper distance between the two world lines is given by the so-called *scale factor*,

$$a(t) \equiv \frac{\Delta\sigma(t)}{\Delta\sigma(t_I)}. \tag{10.11}$$

By using the assumption of isotropy and homogeneity, this scale factor can be shown to be independent of the direction between the two world lines and the spatial location on the hypersurface of homogeneity. Therefore, we can write the line element as

$$ds^2 = -dt^2 + a^2(t)\gamma_{ij}(x^k)dx^i dx^j. \tag{10.12}$$

The factor $\gamma_{ij}(x^k)$ not only describes the geometry of some initial hypersurface, but also the geometry of all other hypersurfaces of homogeneity, as the evolution from one hypersurface to another is completely governed by the scale factor.

Again by assuming isotropy and homogeneity, we can further impose conditions on the factor $\gamma_{ij}(x^k)$. In particular, on surfaces of homogeneity, we can impose an origin and $\gamma_{ij}(x^k)$ must be spherically symmetric around this origin. The factor $\gamma_{ij}(x^k)$ can then be expressed by the most general spherically symmetric form, i.e.

$$\gamma_{ij}(x^k)dx^i dx^j = e^{2\Lambda(r)}dr^2 + r^2\left(d\theta^2 + \sin^2\theta d\phi^2\right), \tag{10.13}$$

where r is some radial distance coordinate, and θ and ϕ are angular coordinates similar to the azimuth and altitude. The Ricci scalar for such a metric is given by

$$R = \frac{2}{r^2}\frac{d}{dr}\left[1 - \left(e^{-2\Lambda}\right)\right]. \tag{10.14}$$

Homogeneity further imposes that the Ricci scalar should be independent of the spatial location. Setting R to some constant k, we can integrate Eq. (10.14) to give

$$\begin{aligned} \gamma_{rr} &= e^{2\Lambda} \\ &= \frac{1}{1 - \frac{1}{6}kr^2 - \frac{A}{r}}, \end{aligned} \tag{10.15}$$

where A is an integration constant. Demanding local flatness, i.e. $\gamma_{rr}(r = 0) = 1$, we have $A = 0$. Finally, redefining k by absorbing the factor of $\frac{1}{6}$ into k, we can write

$$\gamma_{rr} = \frac{1}{1 - kr^2}, \tag{10.16}$$

which completes our description of the hypersurfaces of homogeneity.

10.2.3 Friedmann-Lemaître-Robertson-Walker metric (FLRW)

Plugging Eqs. (10.13) and (10.16) into Eq. (10.12), we arrive at the FLRW metric

$$ds^2 = -dt^2 + a^2(t)\left[\frac{dr^2}{1-kr^2} + r^2\left(d\theta^2 + \sin^2\theta d\phi^2\right)\right]. \qquad (10.17)$$

We can rewrite this metric by defining the coordinate $\chi(r)$ such that

$$d\chi^2 = \frac{dr^2}{1-kr^2}, \qquad (10.18)$$

so that one can also write the FLRW metric as

$$ds^2 = -dt^2 + a^2(t)\left[d\chi^2 + S_k(\chi)^2\left(d\theta^2 + \sin^2\theta d\phi^2\right)\right], \qquad (10.19)$$

where $S_k(\chi)$ is given by

$$S_k(\chi) \equiv \begin{cases} k^{-1/2}\sin(\chi\sqrt{k}) : k > 0 \\ \chi : k = 0 \\ |k|^{-1/2}\sinh(\chi\sqrt{k}) : k < 0. \end{cases} \qquad (10.20)$$

The three cases for k represent three different properties of the spatial hypersurfaces of the Universe. The $k = 0$ case corresponds to a flat Euclidean space, $k > 0$ corresponds to space described by a 3-sphere and finally, $k < 0$ corresponds to space described by a 3-hyperboloid.

This seemingly simple metric will be sufficient to describe the topology of the Universe within the assumptions of isotropy, homogeneity and uniform expansion.

10.2.4 Redshift

When a wave is either moving towards or away from an observer, he/she will see the wavelength to be shorter or longer respectively (Doppler shift). This phenomenon is either referred to as blueshift (approaching source) or redshift (receding source). As the Universe is expanding, we will mostly deal with cosmological redshift, which is defined to be

$$z \equiv \frac{\lambda_{\mathrm{ob}} - \lambda_{\mathrm{em}}}{\lambda_{\mathrm{em}}}, \qquad (10.21)$$

where the subscript 'ob' refers to the observer and the subscript 'em' refers to the emitter. In the comoving coordinate system, the redshift due to the expansion of the Universe is given by

$$z = \frac{a(t_{\mathrm{ob}}) - a(t_{\mathrm{em}})}{a(t_{\mathrm{em}})}. \tag{10.22}$$

Using the FLRW line element in Eq. (10.17), we can relate the redshift to time between successive wave crests. Consider a source at a comoving distance r from an observer. The source emits two successive wave crests at t_{em} and $t_{\mathrm{em}} + \delta t_{\mathrm{em}}$ radially to an observer. This observer will see the wave crests arrive at t_{ob} and $t_{\mathrm{ob}} + \delta t_{\mathrm{ob}}$. Assuming the wave travels at the speed of light, we have $ds^2 = 0$ and therefore

$$\int_{t_{\mathrm{em}}}^{t_{\mathrm{ob}}} \frac{dt}{a(t)} = \int_{\chi_{\mathrm{em}}}^{\chi_{\mathrm{ob}}} d\chi \qquad \text{(first crest)}, \tag{10.23}$$

$$\int_{t_{\mathrm{em}}+\delta t_{\mathrm{em}}}^{t_{\mathrm{ob}}+\delta t_{\mathrm{ob}}} \frac{dt}{a(t)} = \int_{\chi_{\mathrm{em}}}^{\chi_{\mathrm{ob}}} d\chi \qquad \text{(second crest)}. \tag{10.24}$$

Since the comoving distance of the source χ is a constant in the comoving coordinate system, the LHSs of Eqs. (10.23) and (10.24) are equal. This gives us

$$\int_{t_{\mathrm{em}}}^{t_{\mathrm{ob}}} \frac{dt}{a(t)} = \int_{t_{\mathrm{em}}+\delta t_{\mathrm{em}}}^{t_{\mathrm{ob}}+\delta t_{\mathrm{ob}}} \frac{dt}{a(t)}. \tag{10.25}$$

To leading order in δt_{ob} and δt_{em}, we obtain the relationship

$$\begin{aligned} \frac{\delta t_{\mathrm{ob}}}{\delta t_{\mathrm{em}}} &= \frac{a(t_{\mathrm{ob}})}{a(t_{\mathrm{em}})} \\ &= 1 + z. \end{aligned} \tag{10.26}$$

Therefore, according to an observer, a clock at the source appears to be a factor $(1 + z)$ slower, i.e.

$$\delta t_{\mathrm{ob}} = \delta t_{\mathrm{em}} \, (1 + z) \, . \tag{10.27}$$

As a consequence the observed frequency of the source is 'redshifted' according to

$$f_{\mathrm{ob}} = \frac{f_{\mathrm{em}}}{1 + z}, \tag{10.28}$$

and the energy of the wave, given by $E = hf$, is redshifted according to

$$E_{\mathrm{ob}} = \frac{E_{\mathrm{em}}}{1 + z}. \tag{10.29}$$

10.3 Standard Model of Cosmology

With both the content and the topology specified, we are now in the position to use the EFE,

$$G_{\mu\nu} = 8\pi T_{\mu\nu}, \tag{10.30}$$

to link the evolution of the Universe to its content. Using the EFE together with Eqs. (10.19) and (10.1), we can obtain two independent equations relating the scale factor, $a(t)$, to the Universe's content. These two equations are called the *Friedmann equations* and are given by

$$H^2 \equiv \left(\frac{\dot{a}}{a}\right)^2 = \frac{8\pi}{3}\rho_{\text{tot}} + \frac{k}{a^2}, \tag{10.31}$$

$$\frac{\ddot{a}}{a} = -\frac{4\pi}{3}\left(\rho_{\text{tot}} + 3p_{\text{tot}}\right), \tag{10.32}$$

where the overdot denotes the derivative with respect to the co-moving coordinate time. These two equations give us the evolution of the scale factor $a(t)$. They can also be combined to give the 'fluid equation',

$$\dot{\rho}_{\text{tot}} = -3\frac{\dot{a}}{a}\left(\rho_{\text{tot}} + p_{\text{tot}}\right), \tag{10.33}$$

which governs the evolution of the energy density, $\rho(t)$. Before we continue with the evolution of the scale factor, we can have a closer look at the individual energy contributions and their relation to the scale factor.

Matter
The matter content is characterised by its negligible pressure compared to the energy density. The fluid equation for matter then becomes

$$\dot{\rho_M} = -3\frac{\dot{a}}{a}\rho_M. \tag{10.34}$$

This equation has the solution

$$\rho_M = \rho_{m,0}a^{-3}, \tag{10.35}$$

where the subscript 0 denotes the value at the current epoch.

Radiation
For radiation, the pressure is given by $p_R = \frac{1}{3}\rho_R$ and the fluid equation becomes

$$\dot{\rho}_M = -4\frac{\dot{a}}{a}\rho_M, \tag{10.36}$$

Therefore, the evolution of the energy density of radiation is given by

$$\rho_R = \rho_{r,0} a^{-4}. \tag{10.37}$$

Dark Energy
For dark energy, the equation of state is given by $p_{DE} = w\rho_{DE}$. The fluid equation gives

$$\dot{\rho}_{DE} = -3\frac{\dot{a}}{a}\rho_{DE}\left(1 + w(a)\right), \tag{10.38}$$

where $w = w(a)$ can be a general function of the scale factor. The solution to the fluid equation is given by

$$\rho_{DE} = \rho_{DE,0} E(a, w), \tag{10.39}$$

where $E(a, w)$ is given by

$$E(a, w) = e^{-3\int_{a(t_{em})}^{a(t_{ob})} \frac{da'}{a'}(1+w(a'))} \tag{10.40}$$

If the equation of state parameter is $w = -1$, we see that $\rho_{DE} = \rho_{DE,0}$ and the energy density of dark energy is simply a constant throughout the evolution of the Universe. This would be equivalent to having a cosmological constant in the EFE,

$$G_{\mu\nu} = 8\pi T_{\mu\nu} - \Lambda g_{\mu\nu}, \tag{10.41}$$

with the identification that $8\pi\rho_{DE,0} = \Lambda$ [cf. Eq. (10.1)].

In the absence of knowledge about the time dependence of $w(a)$, one can attempt to model it with a series expansion. A commonly used approach [9] is to expand $w(a)$ as a series and terminate the expansion at first order, i.e.

$$\begin{aligned} w(a) &= w_0 + (1 - a)\, w_a + \dots \\ &\simeq w_0 + \frac{z}{1+z} w_a. \end{aligned} \tag{10.42}$$

This expansion has advantages compared to a Taylor expansion around $z = 0$. Firstly, for small z, which is the epoch where we expect dark energy to manifest itself most prominently, this expansion reduces to a Taylor expansion around $z = 0$. However, for large z, the behaviour of this expansion is bounded. With the expansion in Eq. (10.42), we can write $E(a, w)$ as a function of the redshift and two dark energy parameters, w_0 and w_a (setting $z = 0$ at the current epoch), i.e.

$$E(a, w) \approx E(z; w_0, w_a) = (1 + z)^{3(1+w_0+w_a)} e^{-3w_a z/(1+z)}. \tag{10.43}$$

10.3.1 Cosmological Parameters

The total energy density can now be written in terms of the individual components

$$\begin{aligned}\rho_{\text{tot}} &= \sum_i \rho_i \\ &= \rho_M + \rho_R + \rho_{DE} \\ &= \rho_{m,0} a^{-3} + \rho_{r,0} a^{-4} + \rho_{DE,0} E(a, w). \end{aligned} \tag{10.44}$$

Inserting this into the first Friedman equation, Eq. (10.31), we get the following expression for the so-called Hubble parameter,

$$H^2(a) = \frac{8\pi}{3}\left(\rho_{m,0} a^{-3} + \rho_{r,0} a^{-4} + \rho_{DE,0} E(a, w)\right) + \frac{k}{a^2}. \tag{10.45}$$

The Hubble parameter at the current epoch ($a(t_0) = 1$) is thus given by

$$H_0^2 = \frac{8\pi}{3}\left(\rho_{m,0} + \rho_{r,0} + \rho_{DE,0}\right) + k. \tag{10.46}$$

By defining the fractional energy densities

$$\Omega_M \equiv \frac{8\pi \rho_{M,0}}{3H_0^2}, \tag{10.47}$$

$$\Omega_R \equiv \frac{8\pi \rho_{R,0}}{3H_0^2}, \tag{10.48}$$

$$\Omega_{DE} \equiv \frac{8\pi \rho_{DE,0}}{3H_0^2}, \tag{10.49}$$

$$\Omega_k \equiv -\frac{k}{H_0^2}, \tag{10.50}$$

we can write the Hubble parameter as

$$H^2 = H_0^2\left[\Omega_M a^{-3} + \Omega_R a^{-4} + \Omega_k a^{-2} + \Omega_{DE} E(a, w)\right]. \tag{10.51}$$

Evaluating Eq. (10.51) at the current epoch, we obtain the constraint

$$\Omega_M + \Omega_R + \Omega_k + \Omega_{DE} = 1. \tag{10.52}$$

The Hubble parameter can also be written in terms of the redshift as

$$H^2 = H_0^2 \Big[\Omega_M (1+z)^3 + \Omega_R (1+z)^4 + \Omega_k (1+z)^2 \\ + \Omega_{DE} (1+z)^{3(1+w_0+w_a)} e^{-3w_a z/(1+z)} \Big]. \tag{10.53}$$

The scale factor and therefore the evolution of the Universe is determined by a set of cosmological parameters such as $\{H_0, \Omega_M, \Omega_R, \Omega_k, \Omega_{DE}, w_0, w_a\}$.

10.4 Distance and Volume Measures

In Eq. (10.17) we have seen how distances are given as a function of the cosmic scale factor $a(t)$. Furthermore, we have seen in Eq. (10.51) how this scale factor relates to the energy content of the Universe via the Hubble parameter $H \equiv \frac{\dot{a}}{a}$. It is therefore evident that distance, area and volume measures carry with them an imprint of the cosmological parameters. In this section, we aim to make the relationship between distance measures, redshift and the cosmological parameters explicit.

10.4.1 Comoving Distance

The first useful distance measure to define is the *comoving distance*. The comoving distance is the distance between two observers that move along with the cosmological fluid, measured along the null geodesic. The comoving distance therefore does not change due to the expansion of the Universe. The proper distance can be obtained by multiplying the comoving distance by the scale factor $a(t)$. Two types of comoving distances commonly used in cosmology are the radial and the transverse comoving distance.

Radial Comoving Distance

The radial comoving distance, D_C, is obtained by setting $d\theta = d\phi = 0$, and integrating the remaining distance element $d\chi$, i.e.

$$D_C = \int_0^\chi d\chi' \\ = \int_t^0 \frac{dt'}{a(t')} \tag{10.54}$$

We can change the integration to be over the redshift instead of the comoving time by noting that from Eq. (10.26), we can write

$$z(t) = \frac{a(t_0)}{a(t)} - 1, \tag{10.55}$$

where we have set the observer time to be $t_{\text{ob}} = t_0$ and the emitter time to be $t_{\text{em}} = t$. Subsequently, we have

$$dz = -a(t_0)\frac{\dot{a}}{a^2}dt. \tag{10.56}$$

Substituting Eq. (10.56) into Eq. (10.54), we have

$$\begin{aligned} D_C &= -a(t_0)\int_z^0 \frac{a}{\dot{a}}dz' \\ &= a(t_0)\int_0^z \frac{1}{H(z')}dz'. \end{aligned} \tag{10.57}$$

Transverse Comoving Distance

The comoving distance of two objects that are at the same radial comoving distance, but separated on the sky by an angle $\delta\alpha$, is given by $D_M\delta\alpha$. The transverse comoving distance is then given by D_M. Without loss of generality, we can obtain the transverse comoving distance by setting $d\phi = d\chi = 0$. Then, from Eq. (10.19), the transverse comoving distance is given by $S_k(\chi)$, defined in Eq. (10.20). Filling in the appropriate factors for $\chi = D_C$, $k = -\Omega_k H_0^2$ and defining $D_H \equiv H_0^{-1}$, we arrive at

$$D_M = \begin{cases} \frac{D_H}{\sqrt{\Omega_k}} \sinh\left(\frac{D_C}{D_H}\sqrt{\Omega_k}\right) & \text{for } \Omega_k > 0 \\ D_C & \text{for } \Omega_k = 0 \\ \frac{D_H}{\sqrt{|\Omega_k|}} \sin\left(\frac{D_C}{D_H}\sqrt{|\Omega_k|}\right) & \text{for } \Omega_k < 0 \end{cases} \tag{10.58}$$

10.4.2 Luminosity Distance

The luminosity distance is a widely used distance measure in cosmology, because it links together two physical quantities, the bolometric flux F and the bolometric luminosity L. The flux is the measure of the energy received per unit time per unit area and the luminosity gives the total energy emitted by an object, per unit time. The two are related through

$$F = \frac{\mathcal{L}}{4\pi D_L^2}. \tag{10.59}$$

This equation can be seen as the definition of the luminosity distance, i.e.

$$D_L \equiv \sqrt{\frac{\mathcal{L}}{4\pi F}}. \tag{10.60}$$

We can now relate the luminosity distance to the comoving distance. Consider the flux measured by an observer, which is given by

$$F = \frac{dE_{\rm ob}/dt_{\rm ob}}{A}, \tag{10.61}$$

where $dE_{\rm ob}/dt_{\rm ob}$ is the measured luminosity and A is the proper area of the sphere for which all points lie at a fixed proper radial distance. This area A can be found upon inspection of the line element in Eq. (10.19), and is given by

$$\begin{aligned} A &= a^2(t_{\rm ob}) D_M^2 \oiint \sin\theta d\theta d\phi \\ &= 4\pi a^2(t_{\rm ob}) D_M^2. \end{aligned} \tag{10.62}$$

To relate the flux to the intrinsic luminosity, we can use Eqs. (10.27) and (10.29) to write $\frac{dE_{\rm ob}}{dt_{\rm ob}}$ in terms of the emitter's coordinates:

$$\begin{aligned} \frac{dE_{\rm ob}}{dt_{\rm ob}} &= \frac{dE_{\rm em}}{dt_{\rm em}} \frac{1}{(1+z)^2} \\ &= \frac{\mathcal{L}}{(1+z)^2}. \end{aligned} \tag{10.63}$$

Finally, the flux can be written as

$$F = \frac{\mathcal{L}}{4\pi D_M^2 (1+z)^2}. \tag{10.64}$$

Inserting Eq. (10.64) into Eq. (10.60), the luminosity distance is given by

$$\begin{aligned} D_L &= (1+z)\, a(t_{\rm ob}) D_M \\ &= (1+z)\, a(t_{\rm ob}) \begin{cases} \frac{D_H}{\sqrt{\Omega_k}} \sin\left(\frac{D_C}{D_H}\sqrt{\Omega_k}\right) & \text{for } \Omega_k > 0 \\ D_C & \text{for } \Omega_k = 0 \\ \frac{D_H}{\sqrt{|\Omega_k|}} \sinh\left(\frac{D_C}{D_H}\sqrt{|\Omega_k|}\right) & \text{for } \Omega_k < 0. \end{cases} \end{aligned} \tag{10.65}$$

10.4.3 Comoving Volume

Finally, we can also define the *comoving volume*, V_C, in which the number density of objects moving along with the cosmological fluid is constant with respect to the expanding Universe. By inspection of the line element Eq. (10.19), we see that the comoving volume element is given by

$$\begin{aligned} dV_C &= S_k(\chi)^2 d\chi \sin\theta d\theta d\phi \\ &= \frac{D_M^2}{H(z)} dz \sin\theta d\theta d\phi, \end{aligned} \tag{10.66}$$

where we have performed a transformation from χ to z similar to Sect. 10.4.1. This volume element can be integrated over the angular coordinates, and out to redshift z to give the comoving volume [10]

$$V_C = \begin{cases} \frac{4\pi D_H^3}{2\Omega_k}\left[\frac{D_M}{D_H}\sqrt{1+\Omega_k\frac{D_M^2}{D_H^2}} - \Omega_k^{-1/2}\,\text{arcsinh}\left(\sqrt{\Omega_k}\frac{D_M}{D_H}\right)\right] & \text{for } \Omega_k > 0 \\ \frac{4\pi}{3}D_M^3 & \text{for } \Omega_k = 0 \\ \frac{4\pi D_H^3}{2\Omega_k}\left[\frac{D_M}{D_H}\sqrt{1+\Omega_k\frac{D_M^2}{D_H^2}} - |\Omega_k|^{-1/2}\,\text{arcsin}\left(\sqrt{|\Omega_k|}\frac{D_M}{D_H}\right)\right] & \text{for } \Omega_k < 0. \end{cases} \tag{10.67}$$

The distance and volume measures defined above are shown to be functions of the redshift and the cosmological parameters. Therefore, by obtaining distance and redshift information from multiple sources, we infer the cosmological parameters, shown in Sect. 10.3.

Ultimately, we look for the cosmological model that describes the data the best. This could be a spatially flat de Sitter space ($\Omega_M = 0, \Omega_{DE} = 1, \Omega_k = 0, w_0 = -1, w_a = 0$), a spatially flat Einstein-de Sitter model ($\Omega_M = 1$, $\Omega_{DE} = 1$, $\Omega_k = 0$, $w_0 = -1$, $w_a = 0$), or a model with a dark energy that is not in the form of a cosmological constant (Ω_M, Ω_{DE}, Ω_k, w_0, w_a). Despite any possible theoretical preferences, it is the data that will inform us what kind of universe we live in. It is therefore pivotal to have accurate and reliable measurements of the cosmological parameters, so that we can put existing theories in front of empirical tests. In Chap. 11, we will discuss the techniques available to infer the cosmological parameters.

References

1. A.A. Penzias, R.W. Wilson, A measurement of excess antenna temperature at 4080 Mc/s. ApJ **142**, 419–421 (1965)
2. G.F. Smoot et al., Preliminary results from the COBE differential microwave radiometers: large angular scale isotropy of the cosmic microwave background. Astrophys. J. Lett. **371**, L1–L5 (1991)
3. A. Hajian, T. Souradeep, Testing global isotropy of three-year Wilkinson Microwave Anisotropy Probe (WMAP) data: temperature analysis. Phys. Rev. D **74**, 123521 (2006)

4. P. Sarkar et al., The scale of homogeneity of the galaxy distribution in SDSS DR6. Month. Notices Royal Astron. Soc. Lett. **399**(1), L128–L131 (2009)
5. G. Lemaâtre, Un Univers homogéne de masse constante et de rayon croissant rendant compte de la vitesse radiale des nébuleuses extra-galactiques. Annales de la Societe Scietifique de Bruxelles **47**, 49–59 (1927)
6. E. Hubble, A relation between distance and radial velocity among extra-galactic nebulae. Proc. Natl. Acad. Sci. **15**(3), 168–173 (1929)
7. A.G. Riess et al., Observational evidence from supernovae for an accelerating universe and a cosmological constant. Astron. J. **116**(3), 1009 (1998)
8. S. Perlmutter et al., Measurements of and from 42 high-redshift supernovae. Astron. J. **517**(2), 565 (1999)
9. E.V. Linder, Exploring the expansion history of the universe. Phys. Rev. Lett. **90**, 091301 (2003)
10. S.M. Carroll, W.H. Press, E.L. Turner, The cosmological constant. Ann. Rev. Astron. Astrophys. **30**, 499–542 (1992)

Chapter 11
Cosmography

Distance measurements of stellar objects have always been of interest to astronomers and astrophysicists. Nowadays, scientists are interested in distances to these objects because of the potential to inform us about physics on cosmological scales. This was already hinted at in Sect. 10.4, where we showed that the distance as a function of redshift is intimately linked to the energy content and the topology of the Universe. Therefore, measuring the distance and redshift of cosmological object allows one to infer the large scale characteristics of the Universe, also known as *cosmography*.

In Sect. 11.1, we will briefly review the methods currently used to determine distances and redshifts that are based on the measurements of electromagnetic (EM) waves. In Sect. 11.2, we will discuss the potential of GW detections to shed a new light on the cosmological paradigm, by measuring distances and redshifts independently.

11.1 Cosmography Using Electromagnetic Waves

11.1.1 Measuring the Luminosity Distance

Cosmic Distance Ladder

If one were to know the intrinsic luminosity of a distant object, and one could measure its flux here on Earth, one can infer its luminosity distance through Eq. (10.60). Objects that have a known intrinsic luminosity are called *standard candles*. Unfortunately, unambiguous standard candles have yet to be observed. Instead, astronomers use measurements of close-by sources to calibrate the flux of distant objects. This process is repeated to reach increasingly more distant objects. This technique is referred to as the *cosmic distance ladder*, and is the main method to determine distances to astrophysical objects. We will consider a few 'rungs' of the cosmic distance ladder.

T.G.F. Li, *Extracting Physics from Gravitational Waves*,
Springer Theses, DOI 10.1007/978-3-319-19273-4_11

Parallax The oldest method of measuring distance without the aid of some direct ruler measurement is the use of parallax. Parallax is the apparent displacement of an object on the sky as the observer moves. As nearby objects have larger parallax, the distance to an object can be determined.

Astronomers use the concept of parallax to measure distances to nearby stars. In this case, the star's position on the sky compared to distant 'background' stars is measured as the Earth orbits the Sun. Then, using (twice) the distance between the Earth and the Sun, the angle with which the object seems to have moved compared to background stars can be converted into a distance measurement to the object of interest.

Despite this simple concept, parallax plays an important role in calibrating the cosmic distance ladder. Missions such as GAIA are scheduled to collect distance measures of about a billion stars through the parallax method, providing a sturdy first rung of the cosmic distance ladder (e.g. Turon et al. [1]). However, because of the limited distance between Earth and Sun, stellar parallax can only be used up to hundreds of parsecs.

Main Sequence Fitting When the intrinsic luminosity of stars is plotted against their effective temperature, a so-called *Hertzsprung-Russell diagram*, it is found that most stars in their hydrogen-burning phase can often be found along a curve called the *main sequence*. Therefore, measuring the effective temperature of stars through their spectral properties allows one to infer its intrinsic luminosity. Pairing this with the measurement of the flux, one can determine the distance to a star [2–4].

However, in order to construct the Hertzsprung-Russell diagram, we need to somehow know the intrinsic luminosities of a number of stars. One way to do this is to use an alternative technique, such as parallax, to obtain the distance of a star, and use the observed flux to infer the intrinsic luminosity. This is an example of how each rung on the cosmic distance ladder is intertwined with others.

Cepheid Variable Stars Cepheids variables are stars that have a strong relationship between the intrinsic luminosity and the pulsation period [5]. These pulsations are argued to arise from ionised helium acting as a valve for a heat engine, as He^+ is more transparent compared to He^{2+} [6]. As the star expands due to outward radiation pressure, the outer envelope cools down and He^{2+} gets converted to He^+. This conversion will cause the outer layer to be more transparent, further cooling it and causing it to contract. But, due to the contraction, the outer layer heats up again, converting He^+ back into He^{2+}, completing the pulsation cycle.

The strong and measurable relationship makes the Cepheid Variables interesting standard candles, and they are widely used to calibrate other observations [7, 8]. However, the relationship between the pulsation period and the intrinsic luminosity needs calibration itself and depends on assumptions on stellar physics and environmental parameters. The uncertainty in the physics and the phenomenology of Cepheid Variables is still a significant contribution to the error budget of distance measurements.

Type Ia Supernovae Type Ia supernovae (SNIa) are among the most accurate standard candles. They are believed to be produced by a binary system in which a massive star feeds matter onto the accompanying white dwarf [9]. At critical conditions, a violent explosion is triggered in the form of a supernova. The physics behind these events is believed to be roughly independent of the epoch, making them suitable to be standard candles. Although the intrinsic luminosity is not the same for each supernova, there is a phenomenological relationship between the peak luminosity and the amount of fading after 15 days [10]. Exploiting this relationship, SNIa measurements have shown that the Universe appears to be increasingly accelerating. This discovery earned Perlmutter [11], Riess [12] and Schmidt [13] the 2011 Nobel prize in physics.

However, the phenomenological nature of the correlation between the peak luminosity and fading characteristics is a potential seed of errors. For example, as Drell et al. [14] show, a possible temporal evolution in the supernova physics is highly degenerate with effects caused by the expansion of the Universe, adding systematic and statistical uncertainty to any cosmological information obtained from SNIa measurements.

11.1.2 Measuring the Redshift

The measurement of the redshift is also an important tool for astrophysicists to study stellar objects. The emission spectra of stars can provide information about the composition of stars and the dominant photon emission/absorption schemes. Furthermore, measurement of the redshift is crucial to link distance measures to cosmological parameters. The measurement of the redshift is predominantly done through two independent methods: spectroscopy and photometry.

Spectroscopy

The measurement of the redshift through spectroscopy relies on the identification of known spectral lines, such as lines from the Balmer series. The emission of such spectral lines is associated to well understood physical processes and occurs at well defined frequencies. For example, the Balmer spectral lines are a consequence of the $n \geq 3$ to $n = 2$ transitions of a hydrogen atoms, where n represents the principal quantum number. However, the frequencies at which these spectral lines are observed depend on the Doppler effect. Therefore, by comparing the observed frequency of specific spectral lines with their theoretical frequency, the recession velocity, and thus the redshift, can be readily extracted.

As long as the identification of the spectral lines is successful, the redshift can be obtained to accuracies of about $\sigma_z \approx 0.001$ (e.g. Cohen et al. [15]). However, particularly in low signal-to-noise ratios, the absence of clear characteristic spectral lines can lead to erroneous measurement of the redshift. Nevertheless, spectroscopic redshift measurements are widely used and trusted, especially for relatively bright sources.

Photometry

Photometry is an independent way of extracting the redshift by using broad band spectral energy information. The obtained spectral energy distribution is compared with (redshifted) model or empirical spectra to obtain their redshift. Clearly, this photometry relies heavily on understanding the underlying physical processes such as population synthesis and distortion by the interstellar medium (e.g. Abrahamse et al. [16]). The accuracies of such methods are therefore less compared to spectroscopic measurements, $\sigma_z \sim 0.1$ (e.g. Hogg et al. [17]) on top of the potential systematic errors from mismodelling the physical processes.

The determination of the redshift from photometry can also serve as a potent cross-check to spectroscopic measurements. Indeed, photometric measurements have found erroneous redshift measurements by spectroscopic techniques (e.g. Fernandez-Soto et al. [18]). Furthermore, for faint objects without strong spectral features such as distant galaxies or quasars, photometry might be the only channel through which one can measure the redshift.

11.1.3 Direct Measurement of the Cosmological Parameters

There are several methods that allow a direct measurement of the cosmological parameters. Most of these methods require accurate knowledge of the physical processes underlying the measured quantities. Nevertheless, some have proven to provide high quality inference of the cosmological parameters, and others are quickly improving to achieve comparable bounds. Of these techniques, we will only briefly discuss the measurement of the acoustic oscillations in the early Universe, mass density contrast and gravitational lensing.

Acoustic Oscillations in the Early Universe

After the Big Bang, the Universe consisted of hot and dense plasma made up of interacting photons, electrons and protons. In this plasma of photons and matter, small density fluctuations existed due to quantum fluctuations. These fluctuations were subjected to two competing interactions. On one hand, photon pressure drove the dispersion of these fluctuations, and on the other, the gravitational pull tended to restore them. The interplay between these interactions is the cause of the formation of oscillations similar to acoustic oscillations [19, 20].

The most prominent feature of these acoustic oscillations is the so-called *fundamental mode*. This fundamental mode is governed by the distance that a sound wave had travelled until the epoch when photons decouple from matter (epoch of decoupling) ceased its propagation, also known as the *sound horizon*

$$r_s = \int_0^{t_d} c_s dt, \tag{11.1}$$

where c_s is the speed of sound in the baryon-photon plasma and t_d is the epoch of decoupling. The speed of sound depends only on the baryon and radiation content, whereas the time element depends on the entire geometry of the Universe. Therefore, measuring the imprints of these acoustic oscillations can provide us with a direct measurement of the cosmological parameters. Such imprints can be found in the CMB and in the galaxy distribution.

Anisotropies in the Cosmic Microwave Background Acoustic oscillations can be seen as temperature fluctuations in the CMB, among others, on top of the isotropic radiation at a temperature of $T = 2.728$ K and the dipole component due to the motion of the Earth. The angular distribution of these fluctuations encodes information about the sound horizon. In particular, acoustic oscillations cause temperature fluctuations on angular scales corresponding to a half integer number of oscillations within the sound horizon.

Since these angular scales are both a function of the size of the Universe at the epoch of decoupling, and the evolution of the angular distance measure between the epoch of decoupling and the current epoch, cosmological information can be extracted.

Baryon Acoustic Oscillations in Galaxy Distributions As mass overdensities are thought to originate from the fluctuations in the baryon-photon plasma, imprints of the acoustic oscillations are also visible in the galaxy distribution [21–23]. These imprints are called baryon acoustic oscillations (BAO). By comparing the angular scales of the galaxy distribution with the sound horizon from the CMB, one can infer the cosmological parameters.

Mass Density Contrast

Distance measures are not the only observables that can inform us about our cosmology. As mentioned above, initial perturbations in the baryon-photon plasma are likely to be the seeds of large-scale structure formation. Properties such as the probability for a structure to have a given mass at a given redshift is governed by the content and the geometry of the Universe. This is because overdense mass regions tend to increase the densities through gravitational attraction, but the expansion of the Universe tends to decrease such densities.

To quantify this interplay, one looks at the so-called *mass density contrast*

$$\delta = \frac{\Delta \rho_M}{\rho_M} \tag{11.2}$$

where ρ_M is the background matter density and $\Delta \rho_M$ is the matter overdensity around the background density. Then, in linear perturbation theory of GR, the mass density contrast in a matter dominated universe follows the relationship (e.g. Peebles [24])

$$\ddot{\delta} + 2H\dot{\delta} = \frac{3\Omega_M H_0^2}{2a^3}\delta, \tag{11.3}$$

where an overdot represents the time derivative. It can readily be seen that cosmological parameters can be inferred by measuring the mass density contrast δ as a function of the redshift. To probe the mass density contrast, one can measure the galaxy cluster distribution.

Galaxy Cluster Counting Galaxy cluster counting measures the so-called 'mass function' $N(M, z)$, where M is the mass of the cluster and z is the redshift, which is the cluster count per unit comoving volume for a given mass and redshift. This mass function depends on the mass density contrast $\delta(z)$ and the comoving volume element $dV(z)$, which are both functions of the cosmological parameters.

Gravitational Lensing
When a wave passes through the gravitational field of a massive structure (e.g. cluster of galaxies), the appearance of the source will be different. Thus, the phenomenon where a massive structure acts as a lens for transient waves is the so-called 'gravitational lens' effect [25]. The tidal component of the gravitational field can distort the shape of the source, whereas the magnification will change the apparent brightness of the source. A detailed review of gravitational lensing can be found in [26].
Strong Gravitational Lensing If the influence of the gravitational lens is large enough, multiple 'images' can be observed. This is often referred to as "strong gravitational lensing". Strong lensing can provide us with a distance measure by looking at the different arrival times of spectral features of the individual apparent object [27]. Assuming the mass distribution of the lens is known, this time difference will provide us with an absolute distance measure. When this is combined with the redshift of both the source and the lens, the cosmological parameters, mainly H_0, can be measured.

This method depends heavily on one's ability to map the mass distribution of the lens system, which in itself can be a daunting task. But progress has been made over the years, up to a point where studying strongly lensed systems can provide useful bounds on the Hubble constant (e.g. [28, 29]).
Weak Gravitational Lensing When the lensing effect is weaker and only a single image is visible, the effect is referred to as *weak gravitational lensing*. Weak lensing in the context of cosmology is also called *cosmic shear*. From the principle of isotropy, galaxies are intrinsically randomly orientated. However, cosmic shear introduces a net coherent ellipticity to a sample of observed galaxies. When such an imprint can be obtained, the mass distribution of the lens can be inferred.

Using this method, one can obtain the mass density distributions as a function of the redshift, also known as 'tomography', and subsequently probe the cosmological parameters through both the mass density contrast and the comoving volume. Indeed, measurements of weak lensing signatures have placed bounds on the cosmological parameters [30–32].

Just as strong lensing, weak lensing measurements require an accurate knowledge of the mass distribution of the lens. In the case of weak lensing, this is obtained through analysing the coherence of the ellipticity of galaxies. In order to obtain accurate measurements of the cosmic shear, one needs to correlate and average over many galaxies. Provided that large surveys will measure large numbers of galaxies, weak lensing can indeed become a powerful probe of the cosmological parameters.

11.2 Cosmography Using Gravitational Waves

In Sect. 2.6, we hinted that the measurement of CBC signals yields information about the distance between the source and the observer. However, the distance r in Eqs. (1.97), (1.98) and (2.72) is the distance in the harmonic gauge coordinates.

Before we discuss the measurement of distances and redshifts through GW detections, we will first show what happens to the propagation of GWs on cosmological distances.

11.2.1 Propagation of Gravitational Waves on Cosmological Distances

In Sect. 1.3, we looked at the propagation of GWs in a flat space-time. For a FLRW spacetime, Eq. (1.32) no longer describes the propagation of GWs. Instead, we consider the general curved equivalent, given by

$$\Box h_{\mu\nu} = 0, \tag{11.4}$$

where $\Box$ is the curved spacetime d'Alembertian, which in the case of Eq. (11.4) is given by

$$\Box = \frac{1}{\sqrt{-g}}\partial_\mu\left(\sqrt{-g}g^{\mu\nu}\partial_\nu\right). \tag{11.5}$$

To solve Eq. (11.4) in the FLRW metric, it is instructive to introduce the *conformal time*

$$\eta = \int_0^t \frac{dt'}{a(t')} \quad \text{or} \quad d\eta = \frac{dt}{a(t)}. \tag{11.6}$$

Consequently, we can rewrite Eq. (10.19) as

$$ds^2 = a^2(\eta)\left[-d\eta^2 + d\chi^2 + S_k^2(\chi)d\Omega^2\right]. \tag{11.7}$$

Assuming, for simplicity, a flat but expanding spacetime (i.e. $k = 0$), and replacing $r = \chi$ we can express the line element as

$$ds^2 = a^2(\eta)\left[-d\eta^2 + dr^2 + r^2 d\Omega^2\right]. \tag{11.8}$$

We are interested in the evolution of the spherically symmetric solution of the form

$$h_{\mu\nu} = A_{\mu\nu}(\eta, r)/r, \tag{11.9}$$

similar to Eqs. (1.97), (1.98) and (2.72). Therefore,

$$\begin{aligned} 0 &= \partial_\mu \left(\sqrt{-g} g^{\mu\nu} \partial_\nu\right) h_{\alpha\beta} \\ &= \partial_r \left[a^2 r^2 \partial_r h_{\alpha\beta}\right] - \partial_\eta \left[a^2 r^2 \partial_\eta h_{\alpha\beta}\right] \\ &= \partial_r^2 A_{\alpha\beta} - \partial_\eta^2 A_{\alpha\beta} - 2\frac{\partial_\eta a}{a}\partial_\eta A_{\alpha\beta}. \end{aligned} \tag{11.10}$$

Anticipating the solution to fall off with physical distance instead of r, we further look for a solution of the form $A_{\alpha\beta} = B_{\alpha\beta}/a$, so that

$$0 = \partial_r^2 B_{\alpha\beta} - \partial_\eta^2 B_{\alpha\beta} + \frac{\partial_\eta^2 a}{a} B_{\alpha\beta}. \tag{11.11}$$

As most gravitational wave sources that we can observe with the Advanced LIGO-Virgo network are expected to originate from the matter dominated era, we see from Eq. (10.31) that

$$\begin{aligned} \left(\frac{\partial_t^2 a}{a}\right)^2 &\propto \rho_M \\ &\propto a^{-3}, \end{aligned} \tag{11.12}$$

and subsequently

$$a(\eta) \propto \eta^2. \tag{11.13}$$

Therefore, we have $\partial_\eta^2 a/a \sim \eta^{-2}$. Furthermore, we have $\partial_\eta^2 B_{\alpha\beta} \sim \omega^2 B_{\alpha\beta}$, where ω^2 is the characteristic frequency of the GW. For GWs with wavelengths smaller than the Hubble length D_H, we have $\omega^2 B_{\alpha\beta} \gg \eta^{-2} B_{\alpha\beta}$. Therefore, we can write

$$0 \simeq \partial_r^2 B_{\alpha\beta} - \partial_\eta^2 B_{\alpha\beta}. \tag{11.14}$$

This is approximately a wave equation with the solution $B_{\alpha\beta}(r, \eta) = B_{\alpha\beta}(\eta - r)$. The solution therefore becomes

$$h_{\alpha\beta}(r, \eta) \simeq \frac{B_{\alpha\beta}(\eta - r)}{a(\eta) r}. \tag{11.15}$$

Compared to the solution of a static universe, the solution in an expanding universe can therefore be obtained by replacing r with $ra(\eta)$. We can normalise the conformal time so that $\eta = t$ at some time in the present epoch. But as $a(\eta)$ only varies on cosmological timescales, we can set the normalised conformal time to be $\eta = t$ over the entire time-scale of the GW observation. For the same reason, the scale factor $a(\eta)$ can be written as a constant factor $a(t_0)$, where t_0 is the present epoch. Finally, we have

$$h_{\alpha\beta}(r,t) \simeq \frac{B_{\alpha\beta}(t-r)}{a(t_0)r}. \tag{11.16}$$

In Sect. 10.2, we have seen various physical quantities being "redshifted" in an expanding universe. Therefore, it is more natural to express such quantities in the expressions for GWs in terms of their redshifted form. Firstly, by using Eqs. (10.27) and (10.28), the orbital phase, given by Eq. (2.69), becomes

$$\begin{aligned}\int^{t_{\rm em}} dt'_{\rm em}\omega_{\rm em} &= \int^{t_{\rm em}(1+z)} \frac{dt'_{\rm obs}}{1+z}\omega_{\rm obs}(1+z) \\ &= \int^{t_{\rm obs}} dt'_{\rm obs}\omega_{\rm obs},\end{aligned} \tag{11.17}$$

so that the phase does not change when observing the GW in an expanding universe. Secondly, the polarisation independent factor of the amplitude, which is given to leading order in Eqs. (1.97) and (1.98), can be expressed as

$$\begin{aligned}h_c &= \frac{4\mathcal{M}_c^{5/3}\omega_{\rm em}^{2/3}}{a(t_0)r} \\ &= \frac{4\mathcal{M}_c^{5/3}(1+z)^{2/3}\omega_{\rm obs}^{2/3}}{a(t_0)r} \\ &= \frac{4\mathcal{M}_c^{5/3}(1+z)^{5/3}\omega_{\rm obs}^{2/3}}{a(t_0)r(1+z)} \\ &= \frac{4\mathcal{M}_{c,z}^{5/3}\omega_{\rm obs}^{2/3}}{D_L},\end{aligned} \tag{11.18}$$

where we have defined the redshifted chirpmass to be

$$\mathcal{M}_{c,z} \equiv (1+z)\mathcal{M}_c, \tag{11.19}$$

and where we have used the definition of the luminosity distance from Sect. (10.4.2). We can therefore see that for an expanding universe, the expressions keep the same leading amplitude order form provided we make the substitution $r \to D_L$ and $\mathcal{M}_c \to \mathcal{M}_{c,z}$. For higher amplitude orders, the same can be achieved by multiplying all masses by $(1+z)$ and substituting the distance r by the luminosity distance D_L.

11.2.2 Measuring the Luminosity Distance

As early as 1986, Schutz showed that it is possible to determine the Hubble constant from GW observations, by using the fact that GWs from binary systems encode absolute distance information [33]. The polarisation independent factor in Eq. (11.18) depends on the redshifted chirp mass and the luminosity distance. However, the redshifted chirp mass can be determined through the GW phase (see Eq. 2.71). This is in stark contrast with 'conventional' EM distance measurements such as through type Ia supernovae, which depend heavily on poorly understood physics. As GWs bear resemblance to audio signals, binary systems are often referred to as *standard sirens*, by analogy with EM standard candles.

The use of standard sirens to perform cosmography is limited by several effects. Firstly, weak gravitational lensing (cf. Sect. 11.1.3) will distort the measurement of the luminosity distance, obscuring the true loudness of the source. At redshift of $z = 2$, the root mean squared magnification fluctuation can be as large as 20 % [34]. Furthermore, the measurement of the luminosity distance is limited by the detector sensitivity. The largest contribution to the uncertainty in the distance is due to the limited direction and source orientation sensitivity [35]. As interferometric detectors are sensitive to the strain in Eq. (3.1), there is a large correlation between the distance, the sky position and orientation of the binary (cf. Eqs. 2.72 and 3.1), and consequently, the distance is also measured poorly.

However, a network of detectors, by using triangulation, can provide some directional guide and therefore allow for a measurement of the luminosity distance with a reasonable accuracy. Through Fisher matrix calculations for a LIGO/Virgo network, Cutler and Flanagan showed that the fractional error in the luminosity distance is about $<15\,\%$ for $\sim 8\,\%$ of the sources and $<30\,\%$ for $\sim 60\,\%$ of the sources [36]. A more recent study by Nissanke et al. further showed that the luminosity distance is strongly correlated with the inclination angle of the binary, decreasing the sensitivity to both parameters [35]. This luminosity distance-inclination degeneracy can be broken by, for example, partially disentangling the two GW polarisations through measurements with multiple detectors, by using a prior on the inclination angle of the observed binaries, or by observing some auxiliary channel, such as a simultaneous detection of a beamed EM signal. The use of a coincident detection of an EM signal will be further discussed in Sect. 11.2.3.

For a single ET, despite being composed of 3 detectors at a single location, triangulation does not work because of the negligible distance between individual detectors. However, as before, the distance-inclination degeneracy might be broken, if some prior assumptions are made, or if some auxiliary channel is observed.

11.2.3 Measuring the Redshift

Astrophysical systems do not posses the same characteristics to induce clear and consistent spectral lines in the GW spectrum as they do in the EM spectrum. Consequently, the redshift of an event cannot be obtained through the GW equivalent of spectroscopy. Instead, other techniques have to be deployed to obtain the redshift from GW events.

Electromagnetic Counterpart

One way to measure the redshift associated to a GW event is through the identification of an accompanying EM signal. For example, the binary merger of a NS with either a NS or BH is hypothesised to be the progenitor of a short and intense burst of γ-rays, a so-called SGRB [37]. An EM counterpart like a SGRB could provide the necessary redshift information if the host galaxy of the event can be pinpointed. The error on the redshift measurement in this case will be negligible compared to the error on the distance, so that it is often assumed that the redshift is measured conclusively. Furthermore, as SGRBs are believed to be strongly beamed phenomena [38–40], additional information about the inclination angle of the binary can be obtained, breaking the luminosity distance-inclination degeneracy.

Because of the small error in redshift, and the partial breaking of the luminosity distance-inclination degeneracy, the assumption of an EM event accompanying a GW event is commonly used in studies on standard sirens and their ability to infer the cosmological parameters. For example, Nissanke et al. [35] used MCMC methods to show that with a network of advanced detectors and 15 detected standard sirens, H_0 can be constrained to a 5 % accuracy.

For the 3rd generation detector ET, Monte Carlo simulations performed by Sathyaprakash et al. [41] showed that if H_0 is assumed to be accurately known, then Ω_M, Ω_{DE} and a constant w can be constrained to about 10–20 %, despite weak lensing obscuring the distance measurements. Furthermore, Fisher matrix calculations done by the author and his collaborators showed that standard sirens can improve the bounds on the dark energy parameters w_0 and w_a (cf. Eq. 10.42) obtained by the combination of future CMB, BAO and SNIa measurements by 6 and 5 % percent respectively [42].

The assumption that BNS or BHNS mergers are progenitors of SGRBs, and the subsequent use of SGRBs to determine the redshift of a binary merger do suffer from drawbacks. Firstly, despite our increasing knowledge on SGRBs (e.g. [39, 43]), there is no conclusive evidence that binary mergers are indeed progenitors of SGRBs. Secondly, by comparing the SGRB rate with evolutionary calculations of BNS formation, Belczynski et al. [44] found that the fraction of BNS mergers that have the correct phyiscal characteristics to produce an SGRB, can be as low as 10^{-2}. Thirdly, SGRBs are found to be strongly beamed phenomena, with an opening angle of the jet being smaller than ~20 % [45]. Therefore, not all SGRBs are observable here on Earth, decreasing the probability of observing the SGRB in coincidence with the GW signal. Finally, it can occur that the host galaxy is not identified correctly. For an advanced-detector network, the number of galaxies within the sky position-luminosity distance confidence region can vary between a few (high SNR) to a few hundreds (threshold SNR) [35, 46]. Nuttal and Sutton showed that in the advanced detector era, the probability of finding the true host galaxy can range from ~50 to ~90 %, depending on whether the host-galaxy identification is done with a wide or narrow field-of-view telescope.

Galaxy Catalogue
In his original paper on performing cosmography by using GW observations, Schutz proposed the use of a galaxy catalogue, instead of a direct coincident observation of an EM signal [33]. The idea is to identify *all* the galaxies within the sky position-luminosity distance confidence region as a potential host. Through the measurement of multiple standard sirens, potential host galaxies that did not host the event can be statistically ruled out. The use of a galaxy catalogue to obtain redshift information has been worked out by Del Pozzo for a network of Advanced detectors and he shows that even with tens of standard sirens, the Hubble constant can be measured to accuracies in the order of a few percent.

This technique also has several drawbacks. For example, simulations by Bloom et al. [47] and Bulik et al. [48] showed that between 15–20 % of the binary mergers are expected to occur outside dwarf galaxies. If a significant portion of the standard sirens do not originate from within a galaxy, the results could be significantly biased. Furthermore, this technique relies on having a complete galaxy catalogue. The Sloan Digital Sky Survey (SDSS) is assumed to have a complete galaxy coverage up to about $z \sim 0.1$, which is suitable for advanced detectors. However, for third generation detectors, the galaxy catalogues will have to be extended to be complete up to $z \sim 4$. This poses a significant challenge, even for future missions such as the Large Synoptic Survey Telescope [49]. Furthermore, as GWs can be detected at increasingly larger distance due to technological improvements, the number of potential host galaxies grows. The computational cost of having an enormous number of potential host galaxies could prove to be the limiting factor for future GW detectors. Indeed, as ET is expected to make observations at a rate as high as 10^7 year^{-1} for BNS and BHNS systems, and 10^8 year^{-1} for BBH systems, the analysis of each source with many potential host galaxies can become a daunting task.

Neutron Star Mass Distribution
Next, we discuss the use of the NS mass distribution to estimate the redshift, originally proposed by Markovic [50]. In Sect. 11.2.1, it was shown that the mass associated to the GWs detected here on Earth is the redshifted mass $m' = m(1 + z)$. Suppose we know the intrinsic mass distribution of NSs, e.g. through observations of (binary) pulsars. Then, by inferring the redshifted mass, one can obtain direct redshift information.

This idea critically depends on the assumption that the distribution of the intrinsic mass of the NSs in binaries is known. Current observations suggest that the mass for NSs in BNS systems follows a Gaussian distribution with a mean of $\mu \approx 1.35 M_{\odot}$ and a standard deviation of $\sigma \approx 0.05$ [51, 52]. However, an analysis of BNS observations by Schwab, suggest that the mass distribution of NSs in BNS system could be a bimodal distribution [53]. The two modes in this distribution is hypothesised to be associated to two distinct NS formation scenarios.

The validity of this technique is explored by e.g. Chernoff and Finn [54], and Finn [55] for LIGO/Virgo-like detectors, Taylor et al. [56] for an advanced-detector network, and Taylor and Gair [56] for a network of 3rd generation detectors. For an advanced-detector network, Taylor et al. show that, under the assumption of a flat universe ($\Omega_k = 0$), the Hubble constant H_0 can be determined with accuracies of about $\sim$10 %. For a third generation network, Taylor and Gair showed that the dark energy parameters w_0 and w_a can be constrained to about 50 % accuracy.

Although the idea is potent, its heavy dependence on a known and fixed NS mass distribution could severely bias the results. In partiular, we do not know whether our current understanding of the NS mass distribution is obtained with a biased sample of (binary) NSs. Examples of possible biases include a bias toward long-lived systems, and/or a bias towards systems that have an observable EM signature (e.g. pulsars, x-ray binaries). Furthermore, for events to be suitable, one the objects in the binary will have to be unambiguously identified as a NS.

Tidal Deformation of Neutron Stars

In a recent paper, Messenger and Read [57] showed that redshift information can also be obtained directly from the inspiral signal, by considering the influence of the object's internal structure on the waveform. Specifically, the corrections to the GW waveform due to finite size effects are a function of the equation of state of NS matter and the rest masses. A measurement of these corrections would therefore yield information about both the redshifted and the rest mass, from which the redshift can be inferred. Current constraints on the NS equation of state come from the simultaneous measurement of the mass and the radius of NSs [58–61]. However, these measurements are prone to systematic errors as they are heavily dependent on detailed modelling of the radiation mechanisms at the NS surface and absorption in the interstellar medium.

The necessity of knowledge on the NS equation of state could be a major concern when using this technique. However, GW measurements can provide an interesting solution to this problem. Specifically, measurements of the influence of finite size effects in NSs can also be used to probe the NS equation of state. Flanagan and Hinderer [62], and Hinderer et al. [63] show that an advanced-detector network will have difficulties to constrain all but some extreme cases, but ET does have the potential to constrain the NS equation of state. A scenario can be envisaged that a subset of the detections will be used to obtain the NS equation of state, whereas a different subset will be used to obtain redshift information. This technique is yet to be investigated for its potential in cosmography, but Messenger and Read show that the error on the redshift measurement can be as low as 10 % [57].

References

1. C. Turon, X. Luri, E. Masana, Building the cosmic distance scale: from Hipparcos to Gaia. Astrophys. Space Sci. **71** (2012)
2. M.H. Pinsonneault, D.M. Terndrup, R.B.H.J.R. Stauffer, The distances to open clusters from main-sequence fitting. I. New models and a comparison with the properties of the Hyades eclipsing binary VB 22. Astrophys. J. **598**(1), 588 (2003)
3. M.H. Pinsonneault, D.M. Terndrup, R.B.H.J.R. Stauffer, The distances to open clusters as derived from main-sequence fitting. II. Construction of empirically calibrated isochrones. Astrophys. J. **600**(2), 946 (2004)

4. D. An et al., The distances to open clusters from main-sequence fitting. III. Improved accuracy with empirically calibrated isochrones. Astrophys. J. **655**(1), 233 (2007)
5. H.S. Leavitt, 1777 variables in the magellanic clouds. Ann. Harvard College Obs. **60**, 87–108 (1908)
6. S.A. Zhevakin, Physical basis of the pulsation theory of variable stars. ARA&A **1**, 367 (1963)
7. W.L. Freedman et al., Final results from the Hubble space telescope key project to measure the Hubble constant. Astrophys. J. **553**(1), 47 (2001)
8. W.L. Freedman et al., Final results from the Hubble space telescope key project to measure the Hubble constant. Astrophys. J. **553**(1), 47 (2001)
9. K.A. Postnov, L.R. Yungelson, The evolution of compact binary star systems. Rev. Relat. **9**, 6 (2006)
10. M. Hamuy et al., The morphology of type IA supernovae light curves. Astron. J. **112**, 2438 (1996)
11. S. Perlmutter et al., Measurements of and from 42 high-redshift supernovae. Astrophys. J. **517**(2), 565 (1999)
12. A.G. Riess et al., Observational evidence from supernovae for an accelerating universeand a cosmological constant. Astron. J. **116**(3), 1009 (1998)
13. B.P. Schmidt et al., The high-Z supernova search: measuring cosmic deceleration and global curvature of the universe using type ia supernovae. Astrophys. J. **507**(1), 46 (1998)
14. P.S. Drell, T.J. Loredo, I. Wasserman, Type IA supernovae, evolution, and the cosmological constant. Astrophys. J. **530**, 593–617 (2000)
15. J.G. Cohen et al., Caltech faint galaxy redshift survey. X. A redshift survey in the region of the Hubble deep field north. ApJ **538**, 29–52 (2000)
16. A. Abrahamse et al., Characterizing and propagating modeling uncertainties in photometrically derived redshift distributions. Astrophys. J. **734**(36), 36 (2011)
17. D.W. Hogg et al., A blind test of photometric redshift prediction. Astron. J. **115**, 1418–1422 (1998)
18. A. Fernάndez-Soto et al., On the compared accuracy and reliability of spectroscopic and photometric redshift measurements. ApJS **135**, 41–61 (2001)
19. J. Silk, Cosmic black-body radiation and galaxy formation. ApJ **151**, 459 (1968)
20. P.J.E. Peebles, J.T. Yu, Primeval adiabatic perturbation in an expanding universe. Astrophys. J. **162**, 815 (1970)
21. C. Blake, K. Glazebrook, Probing dark energy using baryonic oscillations in the galaxy power spectrum as a cosmological ruler. Astrophys. J. **594**(2), 665 (2003)
22. W. Hu, Z. Haiman, Redshifting rings of power. Phys. Rev. D **68**, 063004 (2003)
23. H.-J. Seo, D.J. Eisenstein, Probing dark energy with baryonic acoustic oscillations from futurelarge galaxy redshift surveys. Astrophys. J. **598**(2), 720 (2003)
24. P.J.E. Peebles, *Principles of Physical Cosmology*, ed. by P.J.E. Peebles, 1993
25. F. Zwicky. Nebulae as gravitational lenses. Phys. Rev. **51**, 290–290 (1937)
26. M. Bartelmann. Gravitational Lensing. ArXiv e-prints (2010), arXiv:1010.3829
27. S. Refsdal, On the possibility of determining Hubbles parameter and the masses of galaxies from the gravitational lens effect. Mon. Not. Royal Astron. Soc. **128**, 307 (1964)
28. R. Fadely et al., Improved constraints on the gravitational lens Q0957+561. II. Strong lensing. ApJS **711**, 246–267 (2010)
29. S.H. Suyu et al. Dissecting the gravitational lens B1608+656. II. Precision measurements of the Hubble constant, spatial curvature, and the dark energy equation of state. Astrophys. J. **711**, 201–221 (2010)
30. R. Massey et al., COSMOS: three-dimensional weak lensing and the growth of structure. Astrophys. J. Suppl. Ser. **172**, 239–253 (2007)
31. L. Fu et al., Very weak lensing in the CFHTLS wide: cosmology from cosmic shearin the linear regime. Astron. Astrophys. **479**, 9–25 (2008)
32. T. Schrabback et al., Evidence of the accelerated expansion of the universe from weak lensing tomography with COSMOS. Astron. Astrophys. **516**(A63), A63 (2010)

33. B.F. Schutz, Determining the Hubble constant from gravitational wave observations. Nature **323**, 310 (1986)
34. M. Bartelmann, P. Schneider, Weak gravitational lensing. Phys. Rep. **340**(4-5), 291–472 (2001)
35. S. Nissanke et al., Exploring short gamma-ray bursts as gravitational-wave standard sirens. Astrophys. J. **725**, 496–541 (2010)
36. C. Cutler, E.E. Flanagan, Gravitational waves from merging compact binaries: how accurately can one extract the binarys parameters from the inspiral waveform? Phys. Rev. D **49**, 2658–2697 (1984)
37. E. Nakar. Short-hard gamma-ray bursts. Phys. Rep. **442**, 166–236 (2007)
38. A.A. Abdo et al., Fermi observations of high-energy gamma-ray emission from GRB 080916C. Science **323**, 1688 (2009)
39. E. Nakar, A. Gal-Yam, D.B. Fox. The local rate and the progenitor life times of short-hard gamma-ray bursts: synthesis and predictions for the laser interferometer gravitational-wave observatory. Astrophys.J. **650**, 281–290 (2006)
40. L. Rezzolla et al., The missing link: merging neutron stars naturally produce jet like structures and can power short gamma-ray bursts. Astrophys. J. Lett. **732**(1), L6 (2011)
41. B.S. Sathyaprakash, B.F. Schutz, C. Van Den Broeck, Cosmography with the Einstein telescope. Class. Quantum Gravity **27**(21), 215006 (2010)
42. W. Zhao et al., Determination of dark energy by the Einstein telescope: comparing with CMB, BAO, and SNIa observations. Phys. Rev. D **83**(2), 023005 (2011)
43. D. Tsang et al. Resonant Shattering of Neutron Star Crusts. Physical Review Letters 108.1, 011102 (2012), p. 011102
44. K. Belczynski et al., The lowest-mass stellar black holes: catastrophic death of neutron stars in gamma-ray bursts. Astrophys. J. Lett. **680**, L129–L132 (2008)
45. D.A. Frail et al., Beaming in gamma-ray bursts: evidence for a standard energy reservoir. ApJ **562**, L55–L58 (2001)
46. W. Del Pozzo, Inference of cosmological parameters from gravitational waves: applications to second generation interferometers. Phys. Rev. D **86**(4), 043011 (2012)
47. J.S. Bloom, S.R. Kulkarni, S.G. Djorgovski, The observed offset distribution of gamma-ray bursts from their host galaxies: a robust clue to the nature of the progenitors. AJ **123**, 1111–1148 (2002)
48. T. Bulik, K. Belczynski, W. Zbijewsk, Distribution of compact object mergers around galaxies. MNRAS **309**, 629–635 (1999)
49. Z. Ivezic et al., LSST: from science drivers to reference design and anticipated data products. ArXiv e-prints (2008), arXiv:0805.2366
50. D. Markovic, Possibility of determining cosmological parameters from measurements of gravitational waves emitted by coalescing, compact binaries. Phys. Rev. D **48**, 4738–4756 (1993)
51. F. Özel et al., On the mass distribution and birth masses of neutron stars. ApJ **757**, 55 (2012)
52. R. Valentim, E. Rangel, J.E. Horvath, On the mass distribution of neutron stars. MNRAS **414**, 1427–1431 (2011)
53. J. Schwab, P. Podsiadlowski, S. Rappaport, Further evidence for the bimodal distribution of neutron-star masses. ApJ **719**, 722–727 (2010)
54. D.F. Chernoff, L.S. Finn, Gravitational radiation, inspiraling binaries, and cosmology. Astrophys. J. Lett. 411, L5–L8 (1993)
55. L.S. Finn, Binary inspiral, gravitational radiation, and cosmology. Phys. Rev. D **53**, 2878–2894 (1996)
56. S.R. Taylor, J.R. Gair, I. Mandel, Cosmology using advanced gravitational wave detectors alone. Phys. Rev. D **85**, 023535 (2012)
57. C. Messenger, J. Read, Measuring a cosmological distance-redshift relationship using only gravitational wave observations of binary neutron star coalescences. Phys. Rev. Lett. **108**, 091101 (2012)
58. F. Özel, Soft equations of state for neutron-star matter ruled out by EXO0748-676. Nature **441**, 1115–1117 (2006)

59. D.A. Leahy, S.M. Morsink, C. Cadeau, Limits on mass and radius for the millisecond-period X-ray pulsar SAX J1808.4-3658. ApJ **672**, 1119–1126 (2008)
60. F. Özel, T. Güver, D. Psaltis, The mass and radius of the neutron star in EXO 1745-248. ApJ **693**, 1775–1779 (2009)
61. D.A. Leahy et al., Constraints on the properties of the neutron star XTE J1814-338 from pulse-shape models. ApJ **691**, 1235–1242 (2009)
62. E.E. Flanagan, T. Hinderer, Constraining neutron-star tidal Love numbers with gravitational-wave detectors. Phys. Rev. D **77**, 021502 (2008)
63. T. Hinderer et al., Tidal deformability of neutron stars with realistic equations of state and their gravitational wave signatures in binary inspiral. Phys. Rev. D **81**(12), 123016 (2010)

Chapter 12
Electromagnetic Counterpart as Redshift Measurement

12.1 Introduction

As mentioned in Sect. 11.2.3, a popular assumption on BNSs and BHNSs is that they are the progenitors of SGRBs. To start our investigation into standard sirens as a cosmological tool, we adopt this assumption as well.

The advantage of an EM counterpart in the form of an SGRB is twofold. Firstly, identifying a SGRB allows for the host galaxy and subsequently its redshift to be determined. Secondly, SGRBs are likely to be strongly beamed phenomena, which allows one to constrain the inclination of the compact binary system, breaking the distance-inclination degeneracy.

For this investigation, we will work with ET (see Sect. 3.1.2) as the 3rd generation detector of our choice. With its expected rate of $\mathcal{O}(10^3 - 10^7)$ BNS and BHNS detections per year [1], Einstein Telescope is likely to give enough sources to perform precision cosmology, even under the assumption of an EM counterpart.

12.2 Implementation

In this investigation, we follow a Monte Carlo approach similar to Sathyaprakash et al. [2]. In particular, we simulate many GW detections according to the predicted rates and distributions, and record the values and errors of luminosity distances and redshifts. Then, the most likely cosmological parameters are inferred through a maximum likelihood analysis of the catalogue of simulated sources. The distributions of measured values for the cosmological parameters are obtained by simulating many catalogues of BNS or BHNS systems.

T.G.F. Li, *Extracting Physics from Gravitational Waves*,
Springer Theses, DOI 10.1007/978-3-319-19273-4_12

12.2.1 Simulated Source Distribution

Source parameters
The neutron star mass distribution was chosen to be uniform in the interval $[1, 2]\,M_\odot$, inspired by the current limits of the neutron star mass [3], and the black hole mass was chosen to be uniform between $[3, 10]\,M_\odot$ [4]. Thc ratio between BNS and BHNS events is taken to be 0.03, the same as predicted for Advanced LIGO-Virgo network [5].

Events were placed uniformly on the sky and in comoving volume. Furthermore, we assume that the sources are distributed according to a local rate function, $R(z)$, per unit source time and unit co-moving volume. The redshift distribution of the sources as observed here on Earth then takes the form

$$\begin{aligned} P(z) &\propto \frac{dV_C}{dz}\frac{R(z)}{1+z} \\ &\propto \frac{4\pi D_M^2(z)}{H(z)}\frac{R(z)}{1+z}, \end{aligned} \tag{12.1}$$

where dV_C is given by Eq. (10.66), D_M is the transverse comoving distance defined in Eq. (10.58), $H(z)$ is the Hubble parameter given in e.g. Eq. (10.51) and the factor of $1 + z$ is introduced to convert the local rate into the rate observed here on Earth. For the rate function, $R(z)$, we take

$$R(z) = \begin{cases} 1+2z & \text{for } z < 1 \\ \frac{3}{4}(5-z) & \text{for } 1 < z < 5 \\ 0 & \text{for } z > 5, \end{cases} \tag{12.2}$$

which is a piece-wise fit created by Cutler and Holz [6], which is in turn based on estimates obtained by using population synthesis models and the cosmic star formation history [7]. The distribution of redshifts z is shown in Fig. 12.1.

By only considering those sources that have an associated SGRB, we can make further assumptions on the inclination, ι, and the polarisation, ψ, of the binary with respect to the observer. Since it is expected that SGRB are strongly beamed [8–10], a coincident observation of the SGRB implies that the binary was orientated nearly face-on, i.e. $\iota \approx 0$. Even if the beam is as wide as 40° (corresponding to a maximal inclination of the orbital plane with respect to the line of sight of the binary, $\iota = 20°$), the resulting area will only cover 3 % of the entire unit sphere, made up by the inclination and the orientation. Averaging the Fisher matrix over the inclination ι and the polarisation ψ with the constrain $\iota < 20°$ will therefore be approximately the same as taking $\iota = 0$. Therefore, we take $\iota = 0$ and the dependence for ψ subsequently drops out of the expression for the beam pattern function [(*cf.* Eqs. (3.6) and (3.7)].

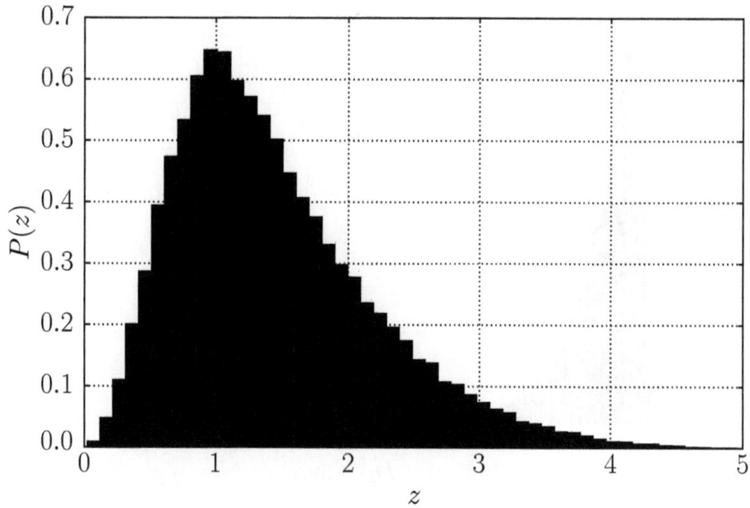

Fig. 12.1 Simulated redshift distribution for GW events from Eq. (12.1)

Number of detectable Sources
ET is expected to detect $\mathcal{O}(10^3-10^7)$ binary neutron star coalescences per year [1]. However, from these, we expect only a small fraction ($\sim 10^{-3}$) to satisfy the constraint that the GW detection coincides with an observation of a SGRB. Assuming a detection rate in the middle of the anticipated range of $\mathcal{O}(10^5)$, we expect to see $\mathcal{O}(10^2)$ events with an associated SGRB. As this number is subject to great uncertainty, we will vary the observed number of sources from 50 up to 10^3 in the results shown in Sect. 12.3.

Cosmological Model
For the simulations we have to choose our 'true' cosmological model. The model was chosen to consist of the set $(H_0, \Omega_M, \Omega_k, \Omega_{DE}, w_0, w_a)$. Because the radiation content is mainly due to the CMB at $T = 2.73$ K, its energy density is $\Omega_r \sim 10^{-5}$ and will be ignored for simplicity.

The values of the cosmological parameters will not be essential in our simulation. What we are interested in is the precision with which they can be measured. However, for consistency with the current measurements of the cosmological parameters, we chose the cosmological parameters to be

$$\begin{aligned} H_0 = 72\ \mathrm{km\,s^{-1}\,Mpc^{-1}}, \quad \Omega_M = 0.27, \quad \Omega_k = 0, \\ \Omega_{DE} = 0.73, \quad w_0 = -1, \quad w_a = 0. \end{aligned} \tag{12.3}$$

Therefore, the challenge is to recover the values in Eq. (12.3) from a set of measured luminosity distances and redshifts. Having defined the 'true' cosmological parameters, the luminosity distance is uniquely determined given a redshift. The dis-

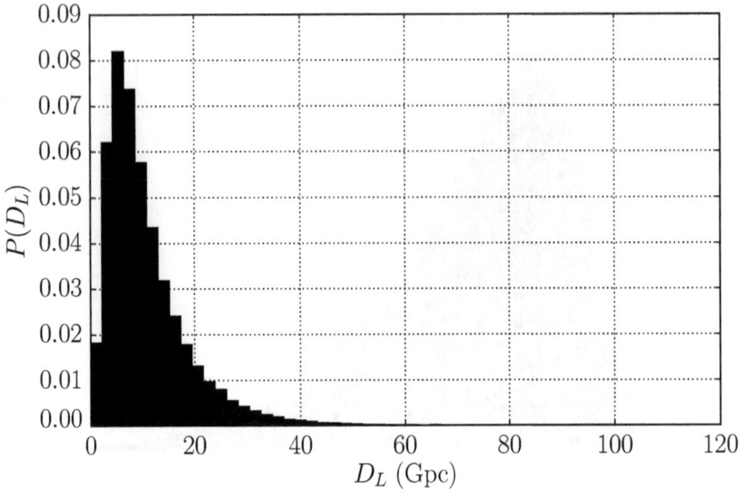

Fig. 12.2 Simulated distribution of luminosity distance given the redshift distribution in Eq. (12.1) and the cosmological parameters in Eq. (12.3)

tribution of luminosity distance given the redshift distribution in Eq. (12.1) is shown in Fig. 12.2.

12.2.2 Simulating the Luminosity Distance Measurement

In line with the SNR threshold currently used at LIGO/Virgo analyses, a GW detection is claimed only when the three ET interferometers have a network SNR of $\rho_{\text{net}} > 8.0$. We will also assume non-spinning CBC systems, which can be characterised by nine parameters $(\mathcal{M}, \eta, t_c, \Phi_c, \theta, \phi, \iota, \psi, D_L)$ (*cf.* Sect. 2.6). For binary neutron stars, the assumption of non-spinning stars is legitimate [11]. However, there is evidence that black holes have significant spin [12]. Considering the low fraction of BHNS systems compared to BNS systems, we will assume spin-less black holes for simplicity. Furthermore, we will only use the leading order in amplitude (see Sect. 2.6). The extra structure in the waveform due to the amplitude correction can improve parameter estimation [13], so we are taking a conservative approach.

With the assumption of associated SGRBs, we can assume the location on the sky, i.e. (θ, ϕ), to be pinpointed by observation of the EM counterpart. Furthermore, in the first ET mock data challenge, Regimbau et al. showed that one can infer the mass parameters to great accuracies [14]. We assume that the mass parameters are accurately determined and do not have considerable correlation with other parameters. This leaves the set of parameters $(t_c, \Phi_c, \iota, \psi, D_L)$ to be estimated. As can be seen from the expression of the waveform in Eq. (2.72), t_c and Φ_c do not appear in the expression of the amplitude. Therefore, we assume that these parameters are largely uncorrelated with the amplitude parameters. Finally, we are left with the set of parameters $\{\iota, \psi, D_L\}$ that we need to consider in order to gauge the ability of ET to estimate the luminosity distance.

Instrumental noise

To estimate the instrumental error on the measurement of the luminosity distance, we turn to the Fisher information matrix (see Sect. 5.2). Suppose that the error on D_L is uncorrelated with the errors on the remaining GW parameters. The Fisher matrix is block diagonal in that case, and one can obtain the Cramer-Rao bound for D_L by immediate inversion of the Fisher information matrix element, i.e.

$$\begin{aligned}\sigma_{D_L}^{\text{inst}} &= \sqrt{C_{D_L D_L}} \\ &= \sqrt{\left(\Gamma^{-1}\right)_{D_L D_L}} \\ &\simeq \sqrt{\left(\Gamma_{D_L D_L}\right)^{-1}} \\ &\simeq \sqrt{\left(\frac{\partial h}{\partial D_L}\middle| \frac{\partial h}{\partial D_L}\right)^{-1}},\end{aligned} \tag{12.4}$$

where h represents the GW waveform. As $h \propto D_L^{-1}$, we have

$$\frac{\partial h}{\partial D_L} = -\frac{h}{D_L}. \tag{12.5}$$

Plugging Eq. (12.5) into Eq. (12.4), the instrumental error of the luminosity distance is given by

$$\sigma_{D_L}^{\text{inst}} \simeq \sqrt{\frac{D_L^2}{(h|\,h)}} \tag{12.6}$$

$$\simeq \frac{D_L}{\rho}, \tag{12.7}$$

where ρ denotes the signal-to-noise ratio. The signal to noise ratio is calculated using the simulated values for GW parameters.

However, we are not in the situation where D_L is uncorrelated with other parameters. In particular, the expression Eq. (12.7) has ignored the correlation with the remaining GW parameters, especially the inclination. To account for this correlation, we note that the maximal effect of the inclination on the SNR is a factor of 2 (between the source being face-on, $\iota = 0$ and edge-on $\iota = \pi/2$). Therefore, to account for the correlation between the D_L and ι, we double the estimate of the error on the luminosity distance, i.e.

$$\sigma_{D_L}^{\text{inst}} \simeq \frac{2D_L}{\rho}. \tag{12.8}$$

Weak lensing
Measurements of the luminosity distance are also affected by weak lensing. The dominant effect is the change of the apparent brightness. This effect can either increase or decrease the apparent brightness, making the inferred luminosity distance to be either smaller or larger. Quantitatively, the effects of the magnification are given by the root mean square of the magnification fluctuation [15]. The magnification fluctuation can be interpreted as an additional error on the measurement of the luminosity distance. Guided by Bartelmann Bartelmann and Schneider [15], we will approximate the error due to weak lensing to be

$$\sigma_{D_L}^{\text{lens}} = 0.09z. \tag{12.9}$$

Total statistical error on the luminosity distance measurement
With the main sources of errors identified above, we can now write the total error on the luminosity distance as

$$\begin{aligned} \sigma_{D_L}^{tot} &= \sqrt{\left(\sigma_{D_L}^{\text{inst}}\right)^2 + \left(\sigma_{D_L}^{\text{lens}}\right)^2} \\ &= \sqrt{\left(\frac{2D_L}{\rho}\right)^2 + (0.09z)^2}. \end{aligned} \tag{12.10}$$

The distribution of luminosity distance errors following the source-parameter distribution given in Sect. 12.2.1 is shown in Fig. 12.3.

Systematic error on the luminosity distance measurement
The final assignment of the measured luminosity distance is done by sampling a Gaussian distribution where the mean is given by the simulated luminosity distance, calculated from the simulated redshift (*cf.* Sect. 12.2.1) and the chosen value of the cosmological parameters, and the standard deviation is set by Eq. (12.10).

12.2.3 Simulating the Redshift Measurement

To simplify the analysis, the redshift measurement by means of EM observations is assumed to be free of any error. This approximation is based on the current errors of spectroscopic redshift determination, which are negligible compared to the errors

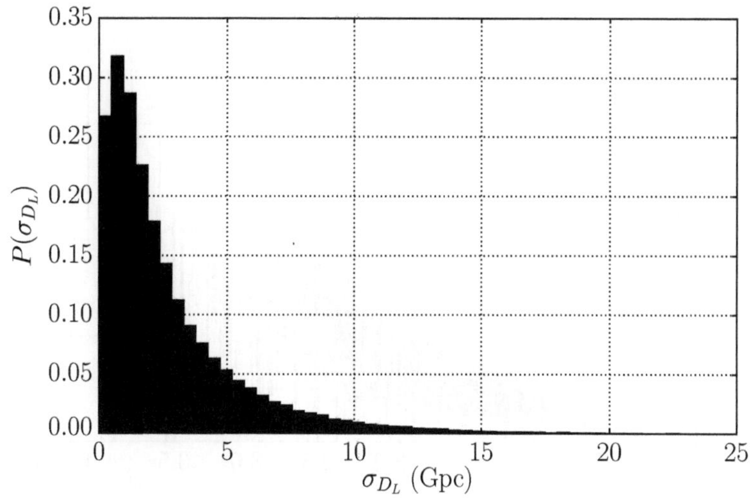

Fig. 12.3 Simulated distribution of luminosity distance errors given the source-parameter distribution given in Sect. 12.2.1

in the luminosity distance (*cf.* Sect. 11.1.2). Although photometric redshift measurements are not as accurate as spectroscopic measurements, significant improvement can be expected by the time ET is operational.

It could also happen that the host-galaxy identification of the EM signal (*e.g.* SGRBs) is incorrect. This could happen in several cases. Firstly, it could happen that the unrelated EM signal occurs at the exact same moment as the GW detection. In this case, the sky location and the redshift can be completely arbitrary. However, considering that there about $\approx$100 SGRBs observed in a year, and the timing resolution is of the order of 1 ms, we do not expect to that the probability of an unrelated coincidence detection to be significant. The other case is that the host galaxy of the SGRB is misidentified. In this case, the sky position is still approximately correct, but the redshift determination might be off. In a study of 20 SGRBs, Bloom et al. [16] showed that for most of these SGRBs, the probability of misidentifying the host galaxy is about 1 %. Therefore, we assume that the correct host galaxy is found for all SGRBs.

12.2.4 Estimating the Cosmological Parameters

The goal of cosmography is to infer the cosmological parameters from a catalogue of observations. In the case presented here, the set of measurements consists of the redshifts, the luminosity distances and the error on the luminosity distance $\left\{\overline{z}^i, \overline{D_L}^i, \overline{\sigma_{D_L}}^i\right\}$. An example catalogue is shown in Fig. 12.4.

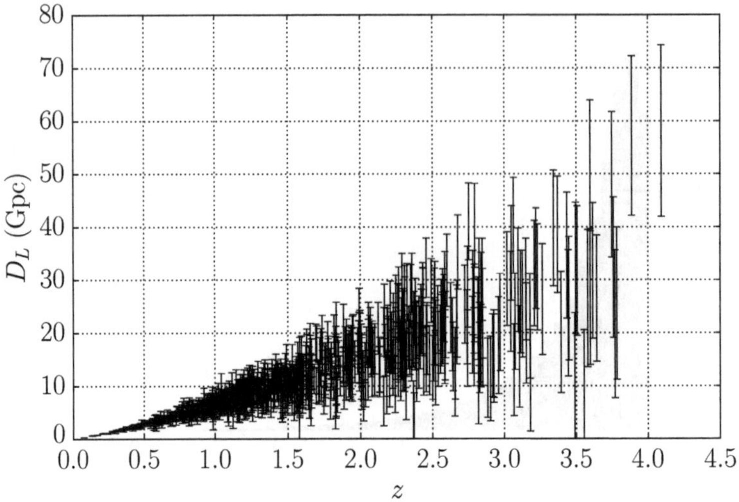

Fig. 12.4 Example catalogue of redshifts, luminosity distances and errors on the luminosity distance

From a set of N observed redshifts, luminosity distances, and the errors on the luminosity distance, we can estimate the cosmological parameters through least-square fitting. Note that least-square fitting is equivalent to finding the maximum of a likelihood that has the shape of a multivariate Gaussian for each of the measurements (*cf.* Sect. 5.1). The residual to minimise in this case is given by

$$\chi^2 = \sum_i^N \left[\frac{\overline{D_L}^i - D_L\left(\overline{z}^i; \vec{\Omega}\right)}{\overline{\sigma_{D_L}}^i} \right]^2, \tag{12.11}$$

where $\vec{\Omega}$ represent the set of cosmological parameters. Finally, we use the Levenberg-Marquardt algorithm (see Sect. 5.1) to minimise the residuals.

However, finding the maximum lilkelihood does not give information on the precision of the inference of the cosmological parameters. To estimate the distribution of cosmological parameters, we perform a Monte Carlo simulation of M realisations. The maximum likelihood for each of these realisations will then give a frequentist approach to the probability of the cosmological parameters.

Measuring all the cosmological parameters at once is a significant challenge, even for the most precise observations we have today. Furthermore, different types of experiments have parameters to which they are more sensitive to, e.g. CMB measurements constrain H_0, Ω_M and Ω_k well, but they have poor accuracy when it comes to dark energy parameters [17]. Therefore, most experiments restrict the analyses to a subset of the cosmological parameters. This amounts to having additional prior information that (strongly) favours specific values for a set of cosmological parameters. In Sect. 12.3 we consider both the cases where some parameters are assumed to be constrained, and the case where standard sirens alone provide the constraints on the cosmological parameters.

12.3 Results

12.3.1 Inferring the Hubble Constant

The earliest measurement of the Hubble constant, H_0, performed by Hubble in 1929, was accomplished by comparing the distances of galaxies with their redshift [18]. The value found was 500 km s^{-1} Mpc^{-1}, which is about ten times bigger than the estimate we have now. With increasing knowledge and technological advances, the value for the Hubble constant has seen a sharp decrease. Current estimates come from an array of different experiments. The analyses of SNIa supernovae gives a value of $H_0 = 74.2 \pm 3.6$ km s^{-1} Mpc^{-1}[19]. The analysis of the CMB anisotropies made by WMAP gives $H_0 = 71.0 \pm 2.5$ km s^{-1} Mpc^{-1} [20]. BAO measurements give $H_0 = 68.2 \pm 2.2$ km s^{-1} Mpc^{-1} [21].

We calculate the median (black dots) and the 95 % confidence interval (black error bars) for H_0 as a function of the number of sources (in steps of 50) for the set of free parameters $\{H_0, \Omega_M, \Omega_{DE}\}$. These results, shown in Fig. 12.5, are compared to similar results from WMAP, BAO and SNIa combined. From Fig. 12.5 we see that around 400 GW events are needed to achieve an accuracy comparable to the WMAP, BAO and SNIa combined results.

12.3.2 Matter Density and Cosmological Constant

Next, we look at the ability of standard sirens to measure the energy densities. Despite tight empirical constraints on the spatial flatness, i.e. $-0.0133 < \Omega_k < 0.0084$ [22], the great promise of standard sirens is to provide an independent measurement. Therefore, we will not assume spatial flatness when investigating the ability of standard sirens to measure the energy densities of the universe. Following the constraint in Eq. (10.52), we take $\Omega_k = 1 - \Omega_M - \Omega_{DE}$.

In Fig. 12.6, we show the 2σ-error ellipses on the (Ω_M, Ω_{DE}) plane (black contours) for the set of free parameters $\{H_0, \Omega_M, \Omega_{DE}\}$. We show contours for 50, 100, 200, 500 and 1000 GW events. The results are compared with the current constraints put forward by the combined measurements of WMAP, BAO and SNIa [23].

Although the error ellipses for a 1000 GW events are bigger than the error ellipses given by the combined measurements of WMAP, BAO and SNIa, Fig. 12.6 shows that the error ellipses are oriented in different directions. This means that GW mea-

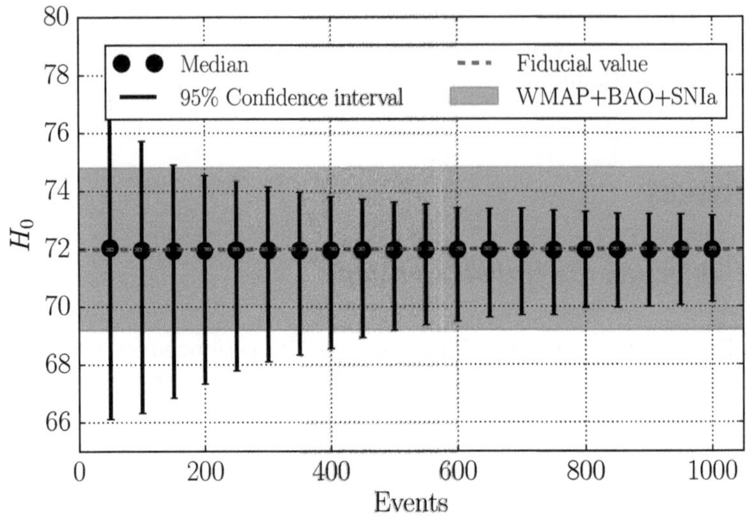

Fig. 12.5 95 % confidence interval (*black lines*) and the median (*black dots*) for H_0 for a variable number of GW events with an EM counterpart. The fiducial value is shown as the *red dashed line*. For comparison, the current bound given by WMAP, BAO and SNIa combined for H_0 is plotted in the *blue shaded area* [22]

surement can cut away some part of the parameter space that has not been constrained yet. In particular, GW measurements are more sensitive to Ω_M.

12.3.3 Dark Energy

Next we want to study the ability of standard sirens to infer the nature of dark energy. As the nature of dark energy is still unknown, we adopt the widely used practice of expanding the equation of state parameter, $w(z)$, as shown in Eq. (10.42). Here, the dark energy equation of state is written in terms of two constants w_0, w_a.

We compare the accuracies on dark energy with optimistic[1] forecasts for the methods introduced in Sects. 11.1.1–11.1.3. The forecasts are based on proposed future missions such as the Square Kilometre Array (SKA) [25], WFIRST (previously known as JDEM) [26] and Euclid [27] which are equipped to study the cosmology through a variety of methods.

Pivot Point

Although we measure two dark energy equation of state parameters, we are mainly interested in the ability to measure the function $w(z)$. As the error on $w(z)$ is itself

[1]For assumptions for the 'optimistic' scenarios of experiments mentioned, consult Albrecht et al. [24].

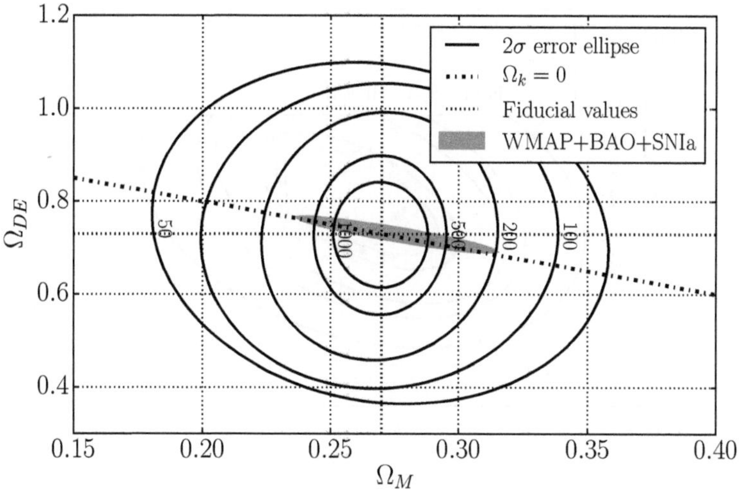

Fig. 12.6 2σ error ellipses in the $\Omega_M - \Omega_{DE}$ plane (*black solid*) for a variable number of GW events considered. These are compared with the current constraints by WMAP, BAO and SNIa (*blue area*) [23]

redshift dependent, it is common practice [28, 29] to make the transform

$$\{w_0, w_a\} \rightarrow \{w_p, w_a\}, \tag{12.12}$$

where w_p is defined to be the value of $w(a)$ at the so-called 'pivot' scale factor a_p, i.e.

$$w_p = w_0 + (1 - a_p) w_a. \tag{12.13}$$

The pivot scale factor is the scale factor at which the error on $w(a)$ is minimal. Following the linearised approximation of $w(a)$, given in Eq. (10.42), the variance of the function $w(a)$ is given by

$$\sigma_w^2 = \sigma_{w_0}^2 + (1 - a)^2 \sigma_{w_a}^2 + 2(1 - a)\text{corr}(w_0, w_a)\sigma_{w_0}\sigma_{w_a}, \tag{12.14}$$

where σ_{w_0} and σ_{w_a} are the standard deviations on w_0 and w_a, and $\text{corr}(w_0, w_a)$ is the correlation between w_0 and w_a. The minimum with respect to the scale factor a is given by

$$a_p = 1 + \frac{\text{corr}(w_0, w_a)\sigma_{w_0}}{\sigma_{w_a}}. \tag{12.15}$$

This expression can then be turned into the pivot redshift, given by

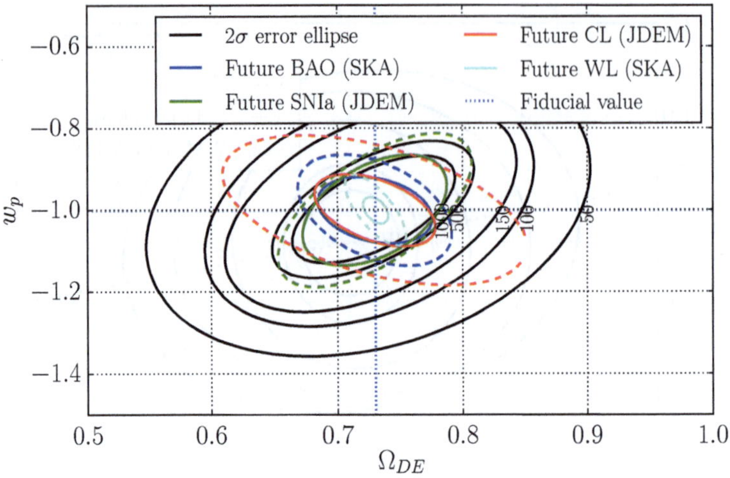

Fig. 12.7 2σ error ellipses on the $\Omega_{DE} - w_p$ plane (*solid, black*). These plots show the potential for standard sirens to constrain the nature of dark energy. The results are compared to the optimistic (*solid, coloured*) and pessimistic (*dashed, coloured*) forecast for BAO, supernovae (*SNIa*), cluster counting (*CL*) and weak lensing (*WL*) for the future Square Kilometre Array or a space-based Joint Dark Energy Mission (*JDEM*) [24]

$$z_p = -\left(1 + \frac{\sigma_{w_a}}{\mathrm{corr}(w_0, w_a)\sigma_{w_0}}\right)^{-1}. \tag{12.16}$$

Besides the equation of state parameters, dark energy is also characterised by its fractional energy density Ω_{DE}. To gauge the ability of an experiment to probe dark energy, it is instructive to visualise the confidence intervals in the $\Omega_{DE} - w_p$ plane. The results for the case where the set $\{\Omega_M, \Omega_{DE}, w_0, w_a\}$ is left free are shown in Fig. 12.7. These results are compared to the optimistic (solid, coloured) and pessimistic (dashed, coloured) predictions for BAO, cluster counting (CL), weak lensing (WL) and supernovae (SNIa) measurements [24].

Figure 12.7 shows that ET can constrain dark energy comparable to future dark energy probes. Assuming that Einstein Telescope observes more than 500 sources, ET even outperforms the optimistic predictions for BAO, supernovae and cluster counting measurements. Only weak lensing has the potential to outperform Einstein Telescope in this case. We can therefore conclude that ET can play a key role in characterising the nature of dark energy.

Dark energy figure of merit

The transformation shown in Eq. (12.12) also has the advantage that the parameters w_p and w_a are uncorrelated. This can be shown as follows. As the transformation in Eq. (12.12) is a linear transformation, the covariance matrix, C_{ij}, is transformed according to

$$C'_{ab} = \sum_{ij} \frac{\partial p_a}{\partial q_i} C_{ij} \frac{\partial p_b}{\partial q_j}. \tag{12.17}$$

Evaluating the partial derivatives as

$$\frac{\partial w_p}{\partial w_0} = 1, \qquad \frac{\partial w_p}{\partial w_a} = 1 - a_p, \qquad \frac{\partial w_a}{\partial w_0} = 0, \tag{12.18}$$

we can evaluate the covariance between w_p and w_0 as

$$\begin{aligned} C_{w_p w_a} &= C_{w_0 w_a} + (1 - a_p) C_{w_a w_a}, \\ &= 0, \end{aligned} \tag{12.19}$$

where we have used the expression for a_p in Eq. (12.15).

Because the parameters w_p and w_a are uncorrelated, the area of the error ellipse is simply proportional to $\sigma_{w_a}\sigma_{w_p}$. Furthermore, as the transformation in Eq. (12.12) is only a change in coordinate bases, the area of the error ellipse in the w_p, w_a plane is the same as the area of the error ellipse in the w_0, w_a plane. In order to gauge the ability of an experiment to probe the dark energy equation of state, it is common practice to introduce a figure of merit given by the reciprocal of this area, i.e. $\left(\sigma_{w_p}\sigma_{w_a}\right)^{-1}$. For this figure of merit, the larger this quantity is, the more accurately we can measure the dark energy equation of state parameters.

The dark energy equation of state figure of merit for ET is shown as a function of the number of sources for various sets of free parameters in Fig. 12.8. These results are compared to the predicted range for the individual BAO, WL, CL and SNIa measurements [24], shown as the shaded area. The upper bound of this shaded area is set by the optimistic predictions from weak lensing measurements, whereas the lower limit is set by the pessimistic predictions from cluster counting measurements. The optimistic and pessimistic predictions for the other methods are between the upper and lower bound. It is evident from Fig. 12.8 that standard sirens can provide constraints on dark energy that are comparable to future (~2020) BAO, WL, CL and SNIa experiments.

12.4 Discussion

Gravitational waves from coalescing binaries directly encode the luminosity distance. However, unlike EM measurements, the redshift is not readily measured. One way to obtain the redshift of a GW event is to observe a coincident EM signal and measure its redshift. Candidate EM signals are SGRB, that are hypothesised to be aftermaths of binary systems with at least one neutron star. An additional advantage of observing an accompanying EM signal is that it also encodes information about the sky location of the GW event, making the inference on the other GW parameters more accurate.

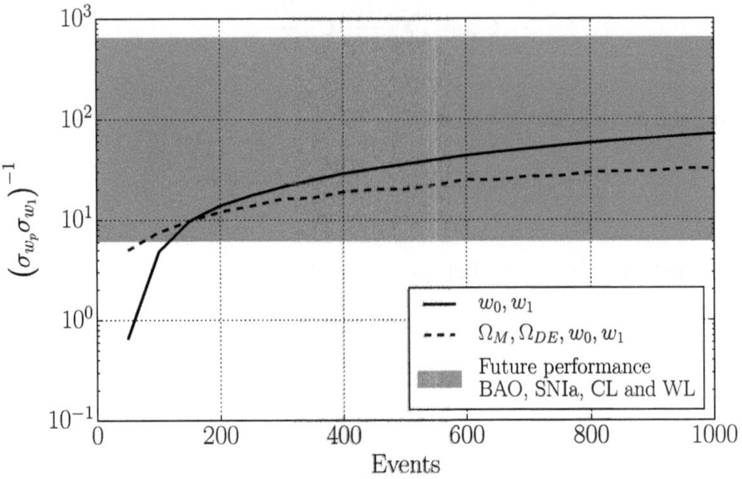

Fig. 12.8 Dark energy figure of merit $(\sigma_{w_p}\sigma_{w_a})^{-1}$ as a function of the number of events for various cases of free parameters. The results are compared future BAO, supernovae (*SNIa*) and cluster counting (*CL*) experiments [24], shown as the shaded area. The upper bound of this *shaded area* is set by the optimistic predictions from weak lensing measurements, whereas the lower limit is set by the pessimistic predictions from cluster counting measurements. Only optimistic predictions for weak lensing experiments have considerably better performances than standard sirens

We investigated the power of ET to infer cosmological parameters by simulating binary systems that have accompanying EM signals. We assumed that the redshift and the sky location are measured without error. We estimated the instrumental error on the luminosity distance by using a Fisher matrix calculation. We also modelled the weak lensing errors linearly in redshift and added this in quadrature to the instrumental error. The redshift and the luminosity distance from a catalogue of sources were then used to infer the cosmological parameters through a maximum likelihood analysis. To obtain statistics on the cosmological parameters, we simulated multiple catalogues of GW events.

In this study, we considered the accuracies on the Hubble constant, the energy content of the Universe, and the dark energy equation of state for various combinations of cosmological parameters to be inferred. For the Hubble constant, ET needs ~400 GW measurements to compete with all current experiments combined. Similar results are obtained for the energy density of matter and dark energy: one needs hundreds of sources to achieve the accuracy of all *current* methods combined. Finally, we investigated the power of ET to probe the dark energy equation of state. In particular, we showed that ET can put constraints on the dark energy and its equation of state comparable to *future* EM measurements, provided that the Hubble constant is known. One way to obtain constraints on the Hubble constant is to use an independent set of close by standard sirens.

The results shown here only consider a single ET, which consist of three co-located GW detectors. On top of the encouraging nature of the results, it should be stressed that it is likely that there will be other 3rd generation detectors by the time ET will be operational. Using multiple detectors at different locations increases the ability to infer the luminosity distance drastically, as shown for various advanced detector network configurations by Veitch et al. [30]. This increased accuracy on the luminosity distance will heavily impact the ability to infer the cosmological parameters. Indeed, using standard sirens to probe the cosmological parameters will provide an independent and complementary alternative to current experiments.

Finally, observing an EM counterpart is not the only option to measure the redshift. As discussed in Sect. 11.2.3, the redshift can also be obtained from the mass distribution of neutron stars, the deformability of neutron stars and galaxy catalogues. In the realistic scenario, these methods will be used in conjunction to improve the inference on the cosmological parameters.

References

1. Einstein Telescope Science Team. Einstein gravitational wave telescope conceptual design study. Technical report ET-0106C-10. 2011
2. B.S. Sathyaprakash, B.F. Schutz, C. Van Den Broeck, Cosmography with the einstein telescope. Class. Quantum Gravity **27**(21), 215006 (2010)
3. F. Ozel et al., On the mass distribution and birth masses of neutron stars. ApJ **757**, 55 (2012)
4. C.L. Fryer, V. Kalogera. Theoretical black hole mass distributions. ApJ. **554**, 548–560 (2001)
5. J. Abadie et al., Calibration of the LIGO gravitational wave detectors in the fifthscience run. Nucl. Instrum. Methods Phys. Res. A **624**, 223–240 (2010)
6. C. Cutler, D.E. Holz, Ultrahigh precision cosmology from gravitational waves. Phys. Rev. D **80**, 104009 (2009)
7. R. Schneider et al., Low-frequency gravitational waves from cosmological compact binaries. Mon. Not. R. Astron. Soc. **324.4**, 797–810 (2001)
8. A.A. Abdo et al., Fermi observations of high-energy gamma-ray emission from GRB 080916C. Science **323**, 1688–1693 (2009)
9. E. Nakar, A. Gal-Yam, D.B. Fox, The local rate and the progenitor lifetimes of short-hard gamma-raybursts: synthesis and predictions for the laser interferometer gravitational-wave observatory. Astrophys. J. **650**, 281–290 (2006)
10. L. Rezzolla et al., The missing link: merging neutron stars naturally produce jetlikestructures and can power short gamma-ray bursts. Astrophys. J. Lett. **732**(1), L6 (2011)
11. R. O'Shaughnessy, C. Kim, Pulsar binary birthrates with spin-opening angle correlations. Astrophys. J. **715**, 230–241 (2010)
12. J.M. Miller et al., Stellar-mass black hole spin constraints from disk reflection andcontinuum modeling. ApJ **697**, 900–912 (2009)
13. C. Van Den Broeck, A.S. Sengupta, Binary black hole spectroscopy. Class. Quantum Gravity **24**(5), 1089 (2007)
14. T. Regimbau et al., Mock data challenge for the einstein gravitational-wave telescope. Phys. Rev. D **86.12**, 122001 (2012)
15. M. Bartelmann, P. Schneider, Weak gravitational lensing. Phys. Rep. **340**(4–5), 291–472 (2001)
16. J.S. Bloom, S.R. Kulkarni, S.G. Djorgovski, The observed offset distribution of gamma-ray bursts from theirhost galaxies: a robust clue to the nature of the progenitors. AJ **123**, 1111–1148 (2002)
17. W. Zhao et al., Determination of dark energy by the einstein telescope: comparingwith cmb, bao, and snia observations. Phys. Rev. D **83**(2), 023005 (2011)

18. E. Hubble, A relation between distance and radial velocity among extra-galacticnebulae. Proc. Natl. Acad. Sci. **15.3**, 168–173 (1929)
19. A.G. Riess et al., A redetermination of the hubble constant with the hubble space telescopefrom a differential distance ladder. Astrophys. J. **699**(1), 539 (2009)
20. N. Jarosik et al., Seven-year wilkinson microwave anisotropy probe (WMAP) observations: sky maps, systematic errors, and basic results. Astrophys. J. Suppl. Ser. **192**(2), 14 (2011)
21. W.J. Percival et al., Baryon acoustic oscillations in the sloan digital sky survey datarelease 7 galaxy sample. Mon. Not. R. Astron. Soc. **401.4**, 2148–2168 (2010)
22. E. Komatsu et al., Seven-year wilkinson microwave anisotropy probe (WMAP) observations: cosmological interpretation. Astrophys. J. Suppl. Ser. **192**, 18 (2011)
23. E. Komatsu et al., Five-year wilkinson microwave anisotropy probe observations: cosmologicalinterpretation. Astrophys. J. Suppl. Ser. **180**(2), 330 (2009)
24. A. Albrecht et al., Report of the dark energy task force. ArXiv Astrophysics e-prints (2006)
25. C.L. Carilli, S. Rawlings, Motivation, key science projects, standards and assumptions. New Astron. Rev. **48**, 979–984 (2004)
26. J. Green et al., Wide-field infrared survey telescope (WFIRST) interim report. ArXiv e-prints (2011), arXiv:1108.1374
27. R. Laureijs, Euclid assessment study report for the ESA cosmic visions. ArXiv e-prints (2009), arXiv:0912.0914
28. D. Huterer, M.S. Turner, Probing dark energy: methods and strategies. Phys. Rev. D **64**(12), 123527 (2001)
29. W. Hu, B. Jain, Joint galaxy-lensing observables and the dark energy. Phys. Rev. D **70.4**, 043009 (2004)
30. J. Veitch et al., Estimating parameters of coalescing compact binaries with proposedadvanced detector networks. Phys. Rev. D **85.10**, 104045 (2012)

Chapter 13
Concluding Remarks

The themes of this work are the questions: "What physical information can be extracted from GWs emitted by CBC systems and how can it be extracted?". To answer these questions, we showed the theoretical models needed to compute CBC waveforms and some of the techniques necessary to extract information from GW detections. We explored two applications in detail: Testing the strong-field dynamics of GR and inferring the large-scale structure of the Universe.

13.1 Testing the Strong-Field Dynamics of General Relativity

In part II, we presented a new theoretical framework called TIGER to test the consistency of the GW signal with the prediction made by GR. This framework makes maximum use of the information provided by applying the concepts of Bayesian hypothesis testing. The novel aspect is the separation of the hypothesis "is the detection consistent with GR?" that is inherently difficult to assess (i.e. one does not know how the signal looks like when GR is not the correct theory), into logically disjoint subhypotheses that are readily assessed. Such a set-up allows for a test of GR without prior knowledge on the specific way in which GR is violated. In other words, TIGER is a *generic* test of GR.

Moreover, we implemented the TIGER framework into a data-analysis pipeline capable of analysing LIGO/Virgo data. We performed comprehensive simulations of BNS detections to test the concepts of the TIGER framework. The results seem to suggest that TIGER can pick up arbitrary deviations from GR, even in the low SNR regime that is applicable to advanced detectors. A comparison with efforts in the literature shows that TIGER is more general and has better performance than existing methods.

For BNS sources, the author and his collaborators are currently expanding TIGER to include waveforms with spin interactions. Furthermore, efforts are ongoing to quantify the effects of possible detector-calibration errors, and to include an

T.G.F. Li, *Extracting Physics from Gravitational Waves*,
Springer Theses, DOI 10.1007/978-3-319-19273-4_13

upper-frequency cut-off to exclude matter effects such as the NS tidal deformability. For BHNS and BBH sources, the author and his collaborators are currently investigating the possibilities to improve the waveform models. In particular, efforts are ongoing to perform numerical simulations of BHNS and BBH systems by using the Tier 1 grid facility at Nikhef.

13.2 Inferring the Large-Scale Structure of the Universe

GWs emitted by CBC systems directly encode the luminosity distance between the source and the observer. The redshift of the source can be inferred from several methods such as using an EM counterpart, a galaxy catalogue, the NS mass distribution, or the NS tidal deformability. With the luminosity distance and redshift determined, one has enough information to use the luminosity distance-redshift relationship to infer the parameters associated with the standard model of cosmology.

In part III, we investigated the ability of ET to infer the cosmological parameters. We considered those CBC sources that can be associated with a redshift measurement from EM counterparts such as SGRBs. For these sources, we calculated the accuracies with which ET can measure the cosmological parameters, and compared them to current and future accuracies from various (purely) EM methods such as the CMB and BAO measurements. Finally, we showed that several hundred sources are needed to achieve accuracies that are similar to EM methods. Since the expected rate for BNS or BHNS detections is $\mathcal{O}(10^3 - 10^7)$ per year, measuring several hundreds of CBC systems that have an associated EM counterpart can be assumed realistic. Therefore, we concluded that future GW observatories such as ET can independently verify the current cosmological paradigm to comparable accuracies as EM methods.

In this book, we have only considered the use of an EM counterpart to provide redshift information. The author and his collaborators are currently investigating the use of the NS tidal deformability to provide redshift information. Preliminary results show that constraints on the cosmological parameters can be set using this method. Ongoing simulations will soon give us an indication of the accuracy with which the cosmological parameters can be measured. Ultimately, all methods of inferring the redshift (EM counterpart, galaxy catalogue, NS mass distribution and NS tidal deformability) should be brought together under a common data-analysis effort. Such a unified data-analysis tool is currently under construction by the author and his collaborators.

13.3 A New Window into the Universe

The two topics presented in this work are merely the tip of the iceberg of what physics can be extracted from direct detections of GWs. For instance, GW measurements yield information about the masses and the spins of BHs and NSs. These can be used to study the stellar, galaxy and BH evolution models. Moreover, CBC and CW signals can shed light on the physical processes that govern the interior of NSs. Bursts of GWs emitted during a supernova can be used to probe its core dynamics. The detection of GWs could lead to observations of speculative compact objects such as cosmic strings, boson stars or sub-solar mass BHs. Furthermore, the association of GWs with an EM transient could inform us about the origin of the transient. For example, the observation of a CBC signal prior to a SGRB will inform us about the physical processes underlying SGRBs.

Although the list of possible physics that can be extracted from GW detections seems to vary across many topics, the procedure to extract the information is remarkably similar. Firstly, one needs to model the expected GWs, similar to what was done in Chaps. 1 and 2 for CBC systems. The difficulty of this stage depends on the system considered. Systems such as CBCs are mostly governed by GR, whereas supernovae require interactions from all corners of physics. Next, these models are used as filters in a detection procedure such as matched filtering (see Chap. 3), where the signal is extracted from the noise. Alternatively, one can use generic filters such as sine-Gaussian filters. In either case, the result is the identification of a GW detection. Finally, once a detection has been made, we can analyse the signal to extract the targeted information. The extraction process can be done with the framework and algorithms associated with Bayesian statistics (see Chaps. 4 and 5). Therefore, the topics covered in this book give insight into the intricate nature of how one arrives at physical statements starting from the raw detector output.

The GW community is entering a critical phase as the upgrades to Advanced LIGO and Advanced Virgo are soon to be completed. If the advanced detectors do not make any detections, serious doubt about progenitor modelling or even the validity of GR might arise. However, if detections are made by the advanced detectors, the possibilities of GW astronomy could be endless. Indeed, the direct detection of GWs could invoke a revolution in astronomy, similar to the one that was driven by the advances of telescopes in the 20th century.

Appendix A
Systematic Multipole Expansion

It is convenient to systematically decompose the multipole expansion of Eq. (1.79) into irreducible representations of the rotation group $SO(3)$. In particular, we will look into a multipolar expansion by using the so-called STF harmonics [1, 2]. Before doing this, it is useful to introduce some notation that will be used throughout the rest of this and next chapter.

A.1 Notation

A tensor F with l (spatial) indices $i_1, i_2, \ldots, i_l$ will be compactly written by using $L = i_1 i_2 \ldots i_l$, i.e.

$$F_L \equiv F_{i_1 i_2 \ldots i_l}. \tag{A.1}$$

Similarly, $F_{L-1} = F_{i_1 i_2 \ldots i_{l-1}}$ and $G_{iL} = G_{i i_1 i_2 \ldots i_l}$. Derivative operators can also be compactly written as $\partial_L = \partial_{i_1} \partial_{i_2} \ldots \partial_{i_l}$. Furthermore, we write the product of vectors as

$$x_L \equiv x_{i_1} x_{i_2} \ldots x_{i_l}, \tag{A.2}$$

and reserve $n_i = x_i / r$ as the unit vector in the radial direction. Repeated upper case indices imply the summation over all l indices, i.e.

$$F_L G_L = \sum_{i_1 \ldots i_l} F_{i_1 \ldots i_l} G_{i_1 \ldots i_l}. \tag{A.3}$$

Also, superscripted integers n surrounded by round brackets denote the nth derivative with respect to the retarded time u, i.e.

$$f^{(n)}(u) \equiv d^n f / du^n. \tag{A.4}$$

T.G.F. Li, *Extracting Physics from Gravitational Waves*,
Springer Theses, DOI 10.1007/978-3-319-19273-4

One can obtain the STF part of a general tensor A in two steps. Firstly, one constructs the symmetric part through

$$A_{(L)} = \frac{1}{l!} \sum_{\pi} A_{\pi\{k_1 \ldots k_l\}}, \tag{A.5}$$

where the summation is over all permutations π of the indices $k_1 \ldots k_l$ and the round brackets surrounding a capitalised index denote the symmetrisation in the included indices. Secondly, one removes all traces from the symmetrised tensor $S_L = A_{(L)}$ by

$$\begin{aligned} \hat{A}_L &= \sum_{n=0}^{[l/2]} (-1)^n \frac{l!(2l-2n-1)!!}{(l-2n)!(2l-1)!!(2n)!!} \delta_{(k_1 k_2} \ldots \delta_{k_{2n-1} k_{2n}} S_{k_{2n+1} \ldots k_l) j_1 j_1 \ldots j_n j_n} \\ &= A_{\langle L \rangle}, \end{aligned} \tag{A.6}$$

where a hat denotes the STF part of a tensor and is equivalent to writing angle brackets surrounding indices. Also, $[l/2]$ denotes the largest integer less than or equal to $l/2$. An STF tensor with l indices has $2l + 1$ independent components and is therefore a representation of dimension $2l + 1$ of the special rotation group $SO(3)$. Furthermore, it can be shown that the set of all STF tensors gives a complete set of representations of $SO(3)$ [3].

A.2 Scalar Fields

Before we look into a systematic multipole expansion for gravitational fields, it is instructive to consider a scalar field ϕ that satisfies the wave equation

$$\Box \phi = -4\pi \rho(t, \vec{x}), \tag{A.7}$$

where the source is localised within some distance d. Outside the source, the general solution to the wave equation is given by

$$\phi(t, \vec{x}) = \sum_{l=0}^{\infty} \frac{(-1)^l}{l!} \partial_L \left[\frac{F_L(u)}{r} \right], \tag{A.8}$$

where F_L is an arbitrary function of the retarded time $u = t - r$. This follows from

$$\Box \left[\frac{f(u)}{r} \right] = 0, \tag{A.9}$$

for an arbitrary function $f(u)$. Therefore, each term in the sum of Eq. (A.8) is a solution to the wave equation. The solution in Eq. (A.8) turns out to be the most general solution to Eq. (A.7) [3].

The wave equation can also be solved by using Green's function. The solution can be written as

$$\phi(t,\vec{x}) = \int d^3\vec{y}\, \frac{\rho(t-|\vec{x}-\vec{y}|,\vec{y})}{|\vec{x}-\vec{y}|}, \tag{A.10}$$

which holds both inside and outside the source. By comparing Eqs. (A.8) to (A.10), one can obtain an expression for the multipole moments F_L in terms of the source $\rho(t,\vec{x})$ [4]

$$F_L(u) = \int d^3\vec{y}\, \hat{y}_L \int_{-1}^{1} dz\, \delta_l(z)\rho(u+z|\vec{y}|,\vec{y}), \tag{A.11}$$

where the hat denotes the STF part shown in Eq. (A.6), and where $\delta_l(z)$ is defined as

$$\delta_l(z) \equiv \frac{(2l+1)!!}{2^{l+1}l!}(1-z^2)^l. \tag{A.12}$$

The integration over z can be expressed as

$$\int_{-1}^{1} dz\, \delta_l(z) f(u+z|\vec{y}|,\vec{y}) = \sum_{k=0}^{\infty} \frac{(2l+1)!!}{2^k k!(2l+2k+1)!!}\left(|\vec{y}|\frac{\partial}{\partial u}\right)^{2k} f(u,\vec{y}), \tag{A.13}$$

from which we can see that the integration over z takes the infinite series of derivatives into account.

A.3 Gravitational Fields

One can treat each component of the gravitational field as a scalar and therefore write the general solution to Eq. (1.31) outside the source as

$$\bar{h}^{00}(t,\vec{x}) = 4\sum_{l=0}^{\infty}\frac{(-1)^l}{l!}\partial_L\left[\frac{F_L(u)}{r}\right], \tag{A.14}$$

$$\bar{h}^{0i}(t,\vec{x}) = 4\sum_{l=0}^{\infty}\frac{(-1)^l}{l!}\partial_L\left[\frac{G_{iL}(u)}{r}\right], \tag{A.15}$$

$$\bar{h}^{ij}(t,\vec{x}) = 4\sum_{l=0}^{\infty}\frac{(-1)^l}{l!}\partial_L\left[\frac{H_{ijL}(u)}{r}\right]. \tag{A.16}$$

However, in this form, Eqs. (A.15) and (A.16) are not STF under the exchange of i with one of the indices in L. Therefore, one must further decompose G_{iL} and H_{ijL} into complete STF tensors by using the steps in Eqs. (A.5) and (A.6). Ultimately, one can write the gravitational field components as

$$\bar{h}^{00}(t,\vec{x}) = 4\sum_{l=0}^{\infty}\frac{(-1)^l}{l!}\partial_L\left[\frac{M_L(u)}{r}\right], \tag{A.17}$$

$$\bar{h}^{0i}(t,\vec{x}) = -4\sum_{l=1}^{\infty}\frac{(-1)^l}{l!}\partial_{L-1}\left[\frac{M^{(1)}_{iL-1}(u)}{r} + \frac{l}{l+1}\epsilon_{iab}\partial_a\left(\frac{S_{bL-1}(u)}{r}\right)\right], \tag{A.18}$$

$$\bar{h}^{ij}(t,\vec{x}) = 4\sum_{l=2}^{\infty}\frac{(-1)^l}{l!}\partial_{L-2}\left[\frac{M^{(2)}_{ijL-2}(u)}{r} + \frac{2l}{l+1}\partial_a\left(\frac{\epsilon_{ab(i}S^{(1)}_{j)bL-2}(u)}{r}\right)\right]. \tag{A.19}$$

Similar to Eq. (1.79), one again has two families of moments, M_L and S_L, describing the gravitational fields. Their explicit forms in terms of the stress-energy tensor are given by

$$M_L(u) = \int d^3\vec{y}\int_{-1}^{1}dz\,\left\{\delta_l(z)\hat{y}_L\sigma - \frac{4(2l+1)\delta_{l+1}(z)}{(l+1)(2l+3)}\hat{y}_{iL}\sigma^{(1)}_i\right.$$
$$\left.+\frac{2(2l+1)\delta_{l+2}(z)}{(l+1)(l+2)(2l+5)}\hat{y}_{ijL}\sigma^{(2)}_{ij}\right\}(u+z|\vec{y}|,\vec{y}), \tag{A.20}$$

$$S_L(u) = \int d^3\vec{y}\int_{-1}^{1}dz\,\epsilon_{ab\langle i_l}\left\{\delta_l(z)\hat{y}_{L-1\rangle a}\sigma_b - \frac{(2l+1)\delta_{l+1}(z)}{(l+2)(2l+3)}\hat{y}_{L-1\rangle ac}\sigma^{(1)}_{bc}\right\}$$
$$(u+z|\vec{y}|,\vec{y}), \tag{A.21}$$

where we have defined

$$\sigma \equiv \left(T^{00}+T^{ii}\right),\quad \sigma_i \equiv T^{0i},\quad \sigma_{ij} \equiv T^{ij}. \tag{A.22}$$

We will return to the use of the STF multipole expansion when we consider GWs in the full, non-linear theory of GR.

References

1. R. Sachs, Gravitational waves in general relativity. VI. The outgoing radiation-condition. Proc. R. Soc. Lond. Ser. A Math. Phys. Sci. **264.1318**, 309–338 (1961)
2. A. Trautmann, F.A.E. Pirani, H. Bondi, *Lectures on General Relativity* (Prentice Hall, Englewood Cliffs, 1965)

3. K.S. Thorne, Multipole expansions of gravitational radiation. Rev. Mod. Phys. **52**, 299–340 (1980)
4. L. Blanchet, T. Damour, Post-Newtonian generation of gravitational waves. Annales de L'Institut Henri Poincare Section Physique Theorique **50**, 377–408 (1989)

Index

T.G.F. Li, *Extracting Physics from Gravitational Waves*,
Springer Theses, DOI 10.1007/978-3-319-19273-4